Table of Contents

PREFACE

On March 28, 2014 the Obama Administration released a key element called for in the President's Climate Action Plan: a Strategy to Reduce Methane Emissions. The strategy summarizes the sources of methane emissions, commits to new steps to cut emissions of this potent greenhouse gas, and outlines the Administration's efforts to improve the measurement of these emissions. The strategy builds on progress to date and takes steps to further cut methane emissions from several sectors, including the oil and natural gas sector.

This technical white paper is one of those steps. The paper, along with four others, focuses on potentially significant sources of methane and volatile organic compounds (VOCs) in the oil and gas sector, covering emissions and mitigation techniques for both pollutants. The Agency is seeking input from independent experts, along with data and technical information from the public. The EPA will use these technical documents to solidify our understanding of these potentially significant sources, which will allow us to fully evaluate the range of options for cost-effectively cutting VOC and methane waste and emissions.

1.0   INTRODUCTION

The oil and natural gas exploration and production industry in the U.S. is highly dynamic and growing rapidly. Consequently, the number of wells in service and the potential for greater air emissions from oil and natural gas sources is also growing. There were an estimated 504,000 producing gas wells in the U.S. in 2011 (U.S. EIA, 2012a), and an estimated 536,000 producing oil wells in the U.S. in 2011 (U.S. EIA, 2012b). It is anticipated that the number of gas and oil wells will continue to increase substantially in the future because of the continued and expanding use of horizontal drilling combined with hydraulic fracturing (referred to here as simply hydraulic fracturing).

Due to the growth of this sector and the potential for increased air emissions, it is important that the U.S. Environmental Protection Agency (EPA) obtain a clear and accurate understanding of emerging data on air emissions and available mitigation techniques. This paper presents the Agency's understanding of emissions and available emissions mitigation techniques from a potentially significant source of emissions in the oil and natural gas sector.

Oil and gas production from unconventional formations such as shale deposits or plays has grown rapidly over the last decade. Oil and natural gas production is projected to steadily increase over the next two decades. Specifically, natural gas development is expected to increase by 44% from 2011 through 2040 (U.S. EIA, 2013b) and crude oil and natural gas liquids are projected to increase by approximately 25% through 2019 (U.S. EIA, 2013b). According to the U.S. Energy Information Administration (EIA), over half of new oil wells drilled co-produce natural gas (U.S. EIA, 2013a).The projected growth is primarily led by the increased development of shale gas, tight gas, and coalbed methane resources utilizing new production technology and techniques such as horizontal drilling and hydraulic fracturing.

Along with the increase in number of wells, the amount of related equipment that has the potential to leak will increase as well. The emissions that occur from leaks are in the form of gasses or evaporated liquids that escape to the atmosphere. Some of the potential leak emissions from these sources include methane and VOCs. The proportion of the different types of air emissions is affected by the composition of the gas in the formation. For example, there tends to be a higher concentration of VOCs in wet gas plays than in dry gas plays.

The emissions data and the mitigation techniques in this paper are based on the onshore natural gas leak emissions that occur from natural gas production, processing, transmission, and storage. However, some of these emissions estimates and mitigation techniques are also applicable to oil wells that co-produce natural gas.

For the purposes of this paper, leaks are defined as VOC and methane emissions that occur at onshore facilities upstream of the natural gas distribution system (i.e., upstream of the city gate). This includes leak emissions from natural gas well pads, oil wells that co-produce natural gas, gathering and boosting stations, gas processing plants, and transmission and storage infrastructure. Potential sources of leak emissions from these sites include agitator seals, compressors seals, connectors, pump diaphragms, flanges, hatches, instruments, meters, open-ended lines, pressure relief devices, pump seals, valves, and improperly controlled liquids storage.[1] For the purposes of this paper, emissions from equipment intended to vent as part of normal operations, such as gas driven pneumatic controllers, are not considered leaks. The definition of leak emissions in this paper was derived by reviewing the various approaches taken in the available literature. Many studies and data sources define leak emissions differently, therefore, in the discussion of these various sources in Section 2 the definition each study uses is compared to the definition presented here.

Leak emissions occur through many types of connection points (e.g., flanges, seals, threaded fittings) or through moving parts of valves, pumps, compressors, and other types of process equipment. Changes in pressure, temperature and mechanical stresses on equipment may eventually cause them to leak. Leak emissions can also occur when connection points are not fitted properly, which causes leaks from points that are not in good contact. Other leaks can occur due to normal operation of equipment, which over time can cause seals and gaskets to

---

[1] Emissions from storage vessels are often required to be controlled by state or federal regulations (e.g., reduced by 95%). Emissions beyond the required level of control from control equipment that is not operating properly, such as leaking vapor recovery units or improperly sized combustors, are considered leaks for the purposes of this white paper.

wear. Weather conditions can also affect the performance of seals and gaskets that are intended to prevent leaks. Lastly, leak emissions can occur from equipment that is not operating correctly, such as storage vessel thief hatches that are left open or separator dump valves that are stuck open.

This document provides a summary of the EPA's understanding of VOC and methane leak emissions at onshore oil and natural gas production, processing, and transmission facilities. This includes available emission data, estimates of VOC and methane emissions and available mitigation techniques. Section 2 of this document describes the EPA's understanding of emissions from leaks at onshore oil and natural gas production, processing, and transmission facilities, and Section 3 discusses available mitigation techniques to reduce emissions from leaks at these facilities. Section 4 summarizes the EPA's understanding based on the information presented in Sections 2 and 3, and Section 5 presents a list of charge questions for reviewers to assist the EPA with obtaining a more comprehensive picture of VOC and methane emissions from leaks and available mitigation techniques.

2.0 OIL AND NATURAL GAS SECTOR LEAKS EMISSIONS DATA AND EMISSIONS ESTIMATES

There are a number of published studies that have estimated leak emissions from the natural gas production and petroleum, processing and transmission sector. These studies have used different methodologies to estimate these emissions, including the use of equipment counts and emission factors, extrapolation of emissions from equipment, and measurement and analysis of leaks. In some cases the studies focus on different portions of the natural gas production and petroleum, processing and transmission and storage sector (e.g., well sites), while others try to account for all leak emissions across the oil and gas sectors. Some of these studies are listed in Table 2-1, along with an indication of the type of information contained in the study (i.e., activity level).

Table 2-1. Summary of Major Sources of Leaks Emissions Information

Name	Affiliation	Year of Report	Activity Factor
Protocol for Equipment Leak Emission Estimates (U.S. EPA, 1995)	U.S. Environmental Protection Agency	1995	None
Methane Emissions from the Natural Gas Industry: Equipment Leaks (GRI/U.S. EPA, 1996)	Gas Research Institute (GRI)/ U.S. Environmental Protection Industry	1996	Nationwide
Greenhouse Gas Reporting Program (U.S. EPA, 2013)	U.S. Environmental Protection Agency	2013	Facility
Inventory of Greenhouse Gas Emissions and Sinks: 1990-2012 (U.S. EPA, 2014)	U.S. Environmental Protection Agency	2014	Regional
Measurements of Methane Emissions at Natural Gas Production Sites in the United States (Allen et al., 2013)	Multiple Affiliations, Academic and Private	2013	Nationwide
City of Fort Worth Natural Gas Air Quality Study, Final Report (ERG, 2011)	City of Fort Worth	2011	Fort Worth, TX
Measurements of Well Pad Emissions in Greeley, CO (Modrak, 2012)	ARCADIS/Sage Environmental Consulting/U.S. Environmental Protection Agency	2012	Colorado
Quantifying Cost-Effectiveness of Systematic Leak Detection and Repair Programs Using Infrared Cameras (CL, 2013)	Carbon Limits	2013	Canada and the U.S.
Mobile Measurement Studies in Colorado, Texas, and Wyoming (Thoma, 2012)	U.S. Environmental Protection Agency	2012 and 2014	Colorado, Texas, and Wyoming
Economic Analysis of Methane Emission Reduction Opportunities in the U.S. Onshore Oil and Natural Gas Industries (ICF International, 2014)	ICF International	2014	Nationwide

Name	Affiliation	Year of Report	Activity Factor
Identification and Evaluation of Opportunities to Reduce Methane Losses at Four Gas Processing Plants (Clearstone, 2002)	Clearstone Engineering, Ltd.	2002	4 gas processing plants
Cost-Effective Directed Inspection and Maintenance Control Opportunities at Five Gas Processing Plants and Upstream Gathering Compressor Stations and Well Sites (Clearstone, 2006)	Clearstone Engineering, Ltd.	2006	5 gas processing plants, 12 well sites, 7 gathering stations

Although methane emissions from oil and natural gas production operations have been measured, analyzed and reported in studies spanning the past few decades, VOC emissions from these operations are not as well represented.

2.1   Protocol for Equipment Leak Emission Estimates (U.S. EPA, 1995)

The EPA protocol provides standard procedures for estimating the total organic compound mass emissions from leaks at oil and natural gas production facilities. The protocol provides four different approaches for estimating leak mass emissions at oil and natural gas production sites. The correlation equations and emission factors were developed from leak data collected from refineries, marketing terminals, oil and gas production operations and synthetic organic chemical manufacturing industry (SOCMI) facilities.

Emission factors and correlations have been developed for the following equipment types: valves, pumps, compressors, pressure relief valves, connectors, flanges, and open-ended lines. An "others" category has also been developed for the petroleum industry. Development of emission factor and correlation equations for the oil and natural gas production facilities were derived from data from six gas plants that were screened by the EPA and the American Petroleum Institute [2] and from leak emission measurement data from 24 oil and natural gas

---

[2] DuBose, D.A., J.I. Steinmetz, and G.E. Harris (Radian Corporation). Frequency of Leak Occurrence and Emission

6

production facilities collected by the American Petroleum Institute.[3,4] The emissions calculated from these emission factors and correlation equations are leak emissions that occur at onshore oil and natural gas production and natural gas processing facilities.

Protocol Leak Estimation Approaches

The protocol document provides four approaches that can be used to estimate mass emissions from leaks.

Average Emission Factor Approach

The first approach involves counting the components by type (e.g., valves, pump seals, connectors, flanges, and open-ended lines) and service (e.g., gas, heavy oil, light oil, and water/oil) at the facility and applying the appropriate average oil and gas production operations emission factors to these counts (see Table 2-4 in U.S. EPA, 1995) to calculate the total organic compound emissions from leaking equipment. There is also an "other" equipment type emission factor that was derived for compressors, diaphragms, drains, dump arms, hatches, instruments, meters, pressure relief valves, polished rods, relief valves, and vents.

Although the average emission factors are in units of kilogram per hour per individual source, it is important to note that these factors are most valid for estimating emissions from a population of equipment (U.S. EPA, 1995). The average factors are not intended to be used for estimating emissions from an individual piece of equipment over a short time period (e.g., 1 hour).

Factors for Natural Gas Liquid Plants. Final Report. Prepared for U.S. Environmental Protection Agency. Research Triangle Park, NC. EMB Report No. 80-FOL-1. July 1982.

[3] Fugitive Hydrocarbon Emissions from Oil and Gas Production Operations, API 4589, Star Environmental, Prepared for American Petroleum Institute, 1993.

[4] Emission Factors for Oil and Gas Production Operations, API 4615, Star Environmental, Prepared for American Petroleum Institute, 1995.

Screening Ranges Approach

The second approach to estimating leak emissions is the screening range approach. This approach is intended primarily to aid in the analysis of old datasets that were collected for older regulations that used 10,000 parts per million by volume (ppmv) as the leak definition. This approach uses the results from EPA Method 21 measurement of leak concentration of components to determine the number of components with a leak greater than or equal to 10,000 parts per million (ppm) and the number of components with a leak less than 10,000 ppm. The estimated emissions are then calculated using the count of components by type, service, and screening value ($\geq$10,000 ppm or <10,000 ppm) at the facility and applying the appropriate average oil and gas production operations emission factors to these counts (see Table 2-8 in U.S. EPA, 1995).

This screening range approach is a better indication of the actual leak rate from individual equipment than the average emission factor approach (U.S. EPA, 1995). However, available data indicate that measured mass emission rates can vary considerably from the rates predicted by use of these screening range emission factors.

EPA Correlation Approach

The third approach is a correlation approach that uses the measured Method 21 screening value (in ppm) for each component and inputs that screening value into correlation equations that calculate the emission rate (see Table 2-10 in U.S. EPA, 1995). This approach offers an additional refinement to estimating emissions from leaks by providing an equation to predict mass emission rate as a function of concentration determined by EPA Method 21 screening for a particular equipment type. Correlations for the petroleum industry apply to refineries, marketing terminals and oil and gas production operations. The petroleum industry correlation equations estimate total organic compound (TOC) emission rates.

The EPA Correlation Approach is preferred when actual screening values (in ppm) are available. Correlations can be used to estimate emissions for the entire range of non-zero screening values, from the highest potential screening value to the screening value that represents the minimum detection limit of the monitoring device. This approach involves entering the non-zero, non-pegged screening value into the correlation equation, which predicts the TOC mass emission rate based on the screening value. Default zero emission rates are used for screening values of zero ppmv and pegged emission rates are used for pegged screening values, where the screening value is beyond the upper limit measured by the portable screening device.

The "default-zero" leak rate is the mass emission rate associated with a screening value of zero. (Note that any screening value that is less than or equal to ambient background concentration is considered a screening value of zero.) The correlations mathematically predict zero emissions for zero screening values. However, data collected by the EPA show this prediction to be incorrect (U.S. EPA, 1995), because mass emissions have been measured from equipment having a screening value of zero. A specific goal when revising the petroleum industry correlations was to collect mass emissions data from equipment that had a screening value of zero. These data were used to determine a default-zero leak rate associated with equipment with zero screening values.

Unit Specific Correlation Approach

The fourth approach calls for developing unit-specific correlations and corresponding mass emission rates. This is done by measuring the screening value in ppm and measuring the mass emission rate by "bagging" the component. A component is bagged by enclosing the component to collect leaking vapors. Measured emission rates from bagged equipment coupled with screening values can be used to develop unit-specific screening value/mass emission rate correlation equations. Unit-specific correlations can provide precise estimates of mass emissions from leaks at the process unit. However, it is recommended that unit-specific correlations are only developed in cases where the existing EPA correlations do not give reasonable mass emission estimates for the process unit (U.S. EPA, 1995).

2.2   GRI/EPA Methane Emissions from the Natural Gas Industry, Volume 8: Equipment Leaks (GRI/U.S. EPA, 1996)

This report provides an estimate of annual methane emissions from leaks from the natural gas production sector using the component method. The component method uses average emission factors for components and the average number of components per facility to estimate the average facility emissions. The average facility emissions were then extrapolated to a national estimate using the number of natural gas production facilities.

The study used two approaches to estimate component emissions for the onshore natural gas production, offshore natural gas production, natural gas processing, natural gas transmission and natural gas storage sectors. The first approach involved screening components using a portable hydrocarbon analyzer and using EPA correlation equations (U.S. EPA, 1995) to estimate the leaking emissions. The EPA correlation equations provide an average leak rate per source using the equipment type (e.g., connectors, flanges, open-ended lines, pumps, valves, other), type of material (e.g., gas, heavy oil, light oil, water/light oil), the leak definition used, and the leak fraction determined by the screening. This approach was used to determine component emission factors for some onshore production sources, natural gas processing and the offshore production sector.

The screening of components involved using a portable instrument to detect leaks around, flanges, valves, and other components by traversing the instrument probe over the entire surface of the component. The components were divided into the following categories:

- Valves (gas/vapor, light liquid, heavy liquid)

- Pump Seals (light liquid, heavy liquid)

- Compressor Seals (gas/vapor)

- Pressure Relief Valves (gas/vapor)

- Connectors, which include flanges and threaded unions (all services)

- Open-Ended Lines (all services)

- Sampling Connections (all services)

All components associated with an equipment source or facility were screened using the procedures specified in EPA Method 21. The maximum measured concentration was recorded using a portable instrument that met the specifications and performance criteria in EPA Method 21. In general, an organic vapor analyzer (OVA) that used a flame ionization detector (FID) was used for conducting the screening measurements. A dilution probe was used to extend the upper range of the instrument from 10,000 to 100,000 ppmv.

The second approach used the GRI Hi-Flow™ (trademark of the Gas Research Institute) sampler or a direct flow measurement to replace data measured using the enclosure method. This method was used to determine emission factors for some of the offshore production sources. The sampler has a high flow rate and generates a flow field around the component that captures the entire leak. As the sample stream passes through the instrument, both the flow rate and the total hydrocarbon (THC) concentration are measured. The mass emission rate can then be determined using these measurements. Offshore leak emissions are not covered in this paper; therefore, the estimates derived from this method will not be discussed further.

For onshore natural gas production, the facilities were broken up into two categories; eastern natural gas production and western gas production to account for regional differences in the methane content of the natural gas. The sources of these leak emissions include gas wells, separators, heaters, dehydrators, metering runs and gathering compressors.

A summary of the average equipment emissions, activity factor, and annual methane emissions for the onshore production sector is presented in Table 2-2. These factors have been used in other reports and studies of methane emissions from the oil and gas sector, including the Inventory of U.S. Greenhouse Gas Emissions and Sinks, which will be discussed in more detail in Section 2.1.4.

As shown in the tables, the study estimated that 15,512 million standard cubic feet per year (MMscf/yr) of methane are emitted as leaks from 271,928 onshore natural gas production wells in the U.S. for the 1992 base year. This converts to approximately 292,930 metric tons (MT) of methane emitted to the atmosphere in the base year.

Table 2-2. GRI/EPA National Annual Emission Estimate for Onshore Natural Gas Production in the United States (1992 Base Year)[a]

Equipment	Average Equipment Methane Emissions (scf/yr)	Activity Factor, Equipment Count	Annual Methane Emissions (MMscf)	Annual Methane Emissions (MT)[b]	90% Confidence Interval
Eastern U.S.					
Gas Well	2,595	129,157	335	6,326	27%
Separator	328	91,670	30.1	568	36%
Heater	5,188	260	1.35	25.5	218%
Dehydrator	7,938	1,047	8.31	157	41%
Meters/Piping	3,289	76,262	251	4,740	109%
Gathering Compressors	4,417	129	0.570	10.8	44%
Eastern U.S. Total			626	11,827	46%
Western U.S.					
Gas Well	13,302	142,771	1,899	35,859	25%
Separator	44,536	74,674	3,326	62,805	69%
Heater	21,066	50,740	1,069	20,186	110%
Dehydrator	33,262	36,777	1,223	23,094	32%
Meters/Piping	19,310	301,180	5,816	109,823	109%
Sm Gathering Compressors[c]	87,334	16,915	1,477	27,895	93%
Lg Gathering Compressors[d]	552,000	96	53.0	1,001	136%
Gathering Stations[e]	1,940,487	12	23.3	440	176%
Western U.S. Total			14,886	281,103	45%
Total			15,512	292,930	-

a - Derived from Tables 5-2 and 5-3 (GRI/U.S. EPA, 1996).
b – Annual methane emissions calculated assuming methane density of 41.63 lb/Mscf.
c – Sm. gathering compressor emission factor does not include compressor seal emissions.
d – Lg. gathering compressor emission factor does not include compressor seal or compressor blowdown emissions.
e – Gathering station emission factor does not include site blowdown line emissions.

The national annual methane emissions from natural gas processing were calculated using published statistics from the Oil and Gas Journal. The 1992 data from the journal listed the total number of natural gas processing plants to be 726. The national methane emissions were calculated using this activity factor and the average facility methane emissions for a natural gas processing plant. The plant methane emissions were calculated using average component counts for gas processing equipment (e.g., valves, connectors, open-ended lines, pressure relief valves, blowdown open-ended lines, compressor seals and miscellaneous). For natural gas processing plants, the average emissions from equipment was estimated to be 2.89 MMscf/yr (18.9 MT). The annual methane emissions from the equipment associated with reciprocating compressors and the equipment associated with centrifugal compressors located at natural gas processing plants were estimated to be 0.538 MMscf/yr (10.2 MT) and 0.031 MMscf/yr (0.585 MT), respectively, in 1992. These methane emissions from the gas processing plant and compressors do not include emissions from starter lines, blowdown lines or compressor seals, which are considered to be vented emissions and not leaks for the purposes of this paper. The ratio of reciprocating and centrifugal compressors located at these plants was based on site visit data from 11 natural gas processing plants. The ratio determined from this data was calculated to be 85% reciprocating and 15% centrifugal. Table 2-3 summarizes the national annual methane emissions from natural gas processing plants, which was estimated to be 3,968 MMscf or 74,921 MT.

Table 2-3. GRI/EPA National Annual Emission Estimate for Natural Gas Processing Plants in the United States (1992 Base Year)[a]

Equipment	Average Facility Methane Emissions (MMscf/yr)	Activity Factor, Number of Plants/ Compressors	Annual Methane Emissions (MMscf)	Annual Methane Emissions (MT)[b]	90% Confidence Interval
Gas Processing Plant[c]	2.40	726	1,744	32,925	27%
Reciprocating Compressors[d]	0.538	4,092	2,201	41,571	36%
Centrifugal Compressors[e]	0.031	726	22.5	425	218%
Total			3,968	74,921	46%

a - Derived from Table 5-5 (GRI/U.S. EPA, 1996).
b – Annual methane emissions calculated assuming methane density of 41.63 lb/Mscf.
c – Gas processing plant emission factor does not include site blowdown emissions.
d – Reciprocating compressor emission factor does not include rod packing, blowdown or starter emissions.
e – Centrifugal compressor emission factor does not include compressor seal, blowdown or starter emissions.

The annual methane emission from transmission compressor stations was calculated using activity data based on statistics by the Federal Energy Regulatory Commission (FERC). The data reported to FERC account for 70% of the total transmission pipeline mileage. The split between reciprocating and turbine compressors was estimated using data from the GRI TRANSDAT database. The average methane emissions from compressor station equipment were estimated to be 3.01 MMscf/yr (56.8 MT) in 1992. The annual methane emissions from the equipment associated with reciprocating compressors and the equipment associated with centrifugal compressors located at transmission stations were estimated to 0.552 MMscf/yr (10.4 MT) and 0.018 MMscf/yr (0.34 MT), respectively. Table 2-4 summarizes the national annual methane leak emissions from natural gas transmission stations, which was estimated to be 50,733 MMscf or 957,999 MT. These methane emissions from the compressor station and compressors do not include emissions from starter lines, blowdown lines or compressor seals, which are considered to be vented emissions and not equipment leaks for the purposes of this paper.

Table 2-4. GRI/EPA National Annual Emission Estimate for Natural Gas Transmission Compressor Stations in the United States (1992 Base Year)[a]

Equipment	Average Facility Methane Emissions (MMscf/yr)	Activity Factor, Number of Stations/ Compressors	Annual Methane Emissions (MMscf)	Annual Methane Emissions (MT)[b]	90% Confidence Interval
Compressor Stations[c]	1.94	1,700	3,298	62,276	103%
Reciprocating Compressors[d]	0.552	6,799	3,753	70,869	68%
Centrifugal Compressors[e]	0.018	681	12.3	231	44%
Total			50,733	957,999	52%

a - Derived from Table 5-6 (GRI/U.S. EPA, 1996).
b – Annual methane emissions calculated assuming methane density of 41.63 lb/Mscf.
c – Compressor station emission factor does not include site blowdown emissions.
d – Reciprocating compressor emission factor does not include rod packing, blowdown or starter emissions.
e – Centrifugal compressor emission factor does not include compressor seal, blowdown or starter emissions.

For natural gas storage facilities, the annual methane emissions were calculated using activity data based on published data in Gas Facts. The number of compressors and injection/withdrawal wells located at natural gas storage facilities were estimated using data collected from site visits to eight facilities. The average methane emissions from natural gas storage facilities were estimated to be 6.80 MMscf/yr (128 MT). The annual average methane emissions from an injection/withdrawal well were estimated to be 0.042 MMscf/yr (0.79 MT). The annual methane emissions from equipment for reciprocating and centrifugal compressors were estimated to be 0.47 MMscf/yr (8.9 MT) and 0.017 MMscf/yr (0.32 MT), respectively, in 1992. The national methane emissions from storage facilities were estimated to be 4,644 MMscf or 87,713 MT and are provided in Table 2-5. These methane emissions from the storage facility and compressors do not include emissions from starter lines, blowdown lines or compressor seals, which are considered to be vented emissions and not leaks for the purposes of this paper.

Table 2-5. GRI/EPA National Annual Emission Estimate for Natural Gas Storage Facilities in the United States (1992 Base Year)[a]

Equipment	Average Facility Emissions (MMscf/yr)	Activity Factor, Number of Facilities/ Compressors	Annual Methane Emissions (MMscf)	Annual Methane Emissions (MT)[b]	90% Confidence Interval
Storage Facilities[c]	6.80	475	3,230	61,004	100
Injection/Withdrawal Wells	0.042	17,999	756	14,275	76
Reciprocating Compressors[d]	0.47	1,396	656	12,390	80
Centrifugal Compressors[e]	0.017	136	2.3	43.7	130
Total			4,644	87,713	57

a - Derived from Table 5-7 (GRI/U.S. EPA, 1996).
b – Annual methane emissions calculated assuming methane density of 41.63 lb/Mscf.
c – Storage facility emission factor does not include site blowdown emissions.
d – Reciprocating compressor emission factor does not include rod packing, blowdown or starter emissions.
e – Centrifugal compressor emission factor does not include compressor seal, blowdown or starter emissions.

2.3   Greenhouse Gas Reporting Program (U.S. EPA, 2013)

In October 2013, the EPA released the 2012 greenhouse gas (GHG) data for Petroleum and Natural Gas Systems[5] collected under the Greenhouse Gas Reporting Program (GHGRP). The GHGRP, which was required by Congress in the FY2008 Consolidated Appropriations Act, requires facilities to report data from large emission sources across a range of industry sectors, as well as suppliers of certain GHGs and products that would emit GHGs if released or combusted.

The GHGRP covers a subset of national emissions, as facilities are required to submit annual reports only if total GHG emissions are 25,000 metric tons carbon dioxide equivalent (CO_2e) or more. Facilities use uniform methods prescribed by the EPA to calculate GHG emissions, such as direct measurement, engineering calculations, or emission factors. In some cases, facilities have a choice of using one of the multiple available calculation methods for an emission source provided.

---

[5] The implementing regulations of the Petroleum and Natural Gas Systems source category of the GHGRP are located at 40 CFR Part 98, subpart W.

Methods for calculating emissions from leaks depend on the industry segment. Facilities in the onshore petroleum and natural gas production segment use population counts and population emission factors for calculating emissions from leaks. Population counts are determined based on either (1) a count of all major equipment (wellheads, separators, meters/piping, compressors, in-line heaters, dehydrators, heater-treaters, and headers) multiplied by average component counts specified in the subpart W regulations, or (2) a count of each component individually for the facility. Emissions are then calculated by multiplying population count by the appropriate population emission factor specified in the subpart W regulations.

Facilities in the onshore gas processing and gas transmission segments use counts of leaking components and leak emission factors for calculating emissions from leaks. The counts of leaking components are identified during an annual leak survey using an optical gas imaging (OGI) instrument, EPA Method 21, infrared (IR) laser beam illuminated instrument, or an acoustic leak detection device. Once the leaking components have been identified and counted, the emissions are calculated by multiplying the count of a specific type of leaking component by the appropriate leak emission factor specified in the subpart W regulations.

For the 2012 reporting year, reported methane emissions from leaks from onshore petroleum and natural gas production were 364,453 MT, onshore natural gas processing were 13,527 MT, and onshore natural gas transmission compression were 15,868 MT.

2.4   Inventory of U.S. Greenhouse Gas Emissions and Sinks: 1990-2012 (U.S. EPA, 2014)

The EPA leads the development of the annual Inventory of U.S. Greenhouse Gas Emissions and Sinks (GHG Inventory). This report tracks total U.S. GHG emissions and removals by source and by economic sector over a time series, beginning with 1990. The U.S. submits the GHG Inventory to the United Nations Framework Convention on Climate Change (UNFCCC) as an annual reporting requirement. The GHG Inventory includes estimates of methane and carbon dioxide for natural gas systems (production through distribution) and petroleum systems (production through refining).

The natural gas production system covers all equipment that process or transport natural gas from oil and gas production sites. (All equipment that process or transport hydrocarbon liquids are covered in the oil systems section of the GHG Inventory.) The natural gas production segment is broken into six regions (North East, Midcontinent, Rocky Mountain, South West, West Coast, and Gulf Coast) and includes estimates for gas wells, separation equipment, gathering compressors, gathering pipelines, drilling and well completions, normal operations, condensate tank vents, well workovers, liquids unloading, vessel blowdowns, and process upsets.

For the natural gas production segment, only methane emissions from gas wells, field separation equipment, and gathering compressor systems will be discussed from the GHG Inventory. Leaks from gas wells include emissions from various components, such as connectors and valves, on a wellhead. Field separation equipment includes heaters, separators, dehydrators, meters and piping. Gathering compressor systems include reciprocating compressors, equipment such as scrubbers and coolers associated with the compressors, and the piping. Leaks from field separation equipment and gathering compressor systems include emissions from components in these equipment and systems. The only exception is the gathering compressors source that includes both leak emissions and vented emissions from compressor seals in the GHG Inventory. (Note: Vented emissions from compressors are not defined as leaks in this paper, but are discussed in the white paper on compressors.) The 2014 GHG Inventory (published in 2014; containing emissions data for 1990-2012) calculates potential[6] methane leak emissions from gas wells and field separation equipment using emission factors from the GRI/EPA study (GRI/U.S. EPA, 1996). The emission factors from the GRI/EPA study are split regionally into Eastern and Western factors. These emission factors are adapted in the 2014 GHG Inventory for each of the NEMS regions by adjusting the GRI/EPA emission factors for the NEMS region-specific methane content in produced natural gas. All of the emission factors from the GRI/EPA study assume methane content of 78.8% in the produced natural gas. However, the 2014 GHG

---

[6] The calculation of emissions for each source of in the GHG Inventory generally involves first the calculation of potential emissions (methane that would be emitted in the absence of controls), then the compilation of emissions reductions data, and finally the calculation of net emissions by deducting the reductions data from the calculated potential emissions. This approach was developed to ensure an accurate time series that reflects real emission trends. Key data on emissions from many sources are from GRI/U.S. EPA 1996, and since the time of this study practices and technologies have changed. While the study still represents best available data for some emission sources, using these emission factors alone to represent actual emissions without adjusting for emissions controls would in many cases overestimate emissions. As updated emission factors reflecting changing practices are not available for most sources, the GRI/U.S. EPA 1996 emission factors continue to be used for many sources for all years of the GHG Inventory, but they are considered to be potential emissions factors, representing what emissions would be if practices and technologies had not changed over time.

Inventory uses regional methane contents obtained from a 2001 study by the Gas Technology Institute (GTI) on unconventional gas and gas composition[7] to adjust the GRI/EPA emission factors to account for the regional methane content differences. The GHG Inventory emissions are then calculated by applying the modified GRI/EPA emission factors to component counts for each year of the GHG Inventory. Because component counts are not available for each year of the GHG Inventory, a set of industry activity data drivers was developed and used to update activity data.[8] The 2014 GHG Inventory, emission factors, and methane emissions are presented by region in Table 2-6. The 2014 GHG Inventory estimated 332,662 MT of potential methane leak emissions from gas wells and field separation equipment from natural gas production activities in 2012.

Table 2-6. 2011 Data and Calculated Methane Potential Leak Emissions for the Natural Gas Production Segment by Region[a]

Region	Activity	Activity Data	Emission Factor	Calculated Potential Emissions (MT)
North East	Associated Gas Wells	38,770	NA	NA
	Non-associated Gas Wells	112,607	7.67 scfd/well	6,071
	Gas Wells with Hydraulic Fracturing	46,367	7.54 scfd/well	2,457
	Heaters	318	15.38 scfd/heater	34
	Separators	112,872	0.97 scfd/separator	771
	Dehydrators	22,164	23.53 scfd dehydrator	3,665
	Meters/Piping	7,910	9.75 scfd/meter	542

[7] GRI-01/0136 GTI's Gas Resource Database: Unconventional Natural Gas and Gas Composition Databases. Second Edition. August, 2001.
[8] For example, recent data on various types of field separation equipment in the production stage (i.e., heaters, separators, and dehydrators) are unavailable. Each of these types of field separation equipment was determined to relate to the number of gas wells. Using the number of each type of field separation equipment estimated by GRI/EPA in 1992, and the number of gas wells in 1992, a factor was developed that is used to estimate the number of each type of field separation equipment throughout the time series. The annual well count data used for these sources were obtained from a production database maintained by DrillingInfo, Inc. (DrillingInfo, 2012).

Region	Activity	Activity Data	Emission Factor	Calculated Potential Emissions (MT)
Midcontinent	Associated Gas Wells	27,470	NA	NA
	Non-associated Gas Wells	77,896	7.45 scfd/well	4,080
	Gas Wells with Hydraulic Fracturing	30,156	8.35 scfd/well	1,771
	Heaters	43,869	14.9 scfd/heater	4,596
	Separators	47,003	0.94 scfd/separator	311
	Dehydrators	15,064	95.54 scfd dehydrator	10,118
	Meters/Piping	143,186	9.45 scfd/meter	9,509
Rocky Mountain	Associated Gas Wells	32,598	NA	NA
	Non-associated Gas Wells	9,665	35.05 scfd/well	2,381
	Gas Wells with Hydraulic Fracturing	73,755	40.72 scfd/well	21,115
	Heaters	38,040	56.73 scfd/heater	15,172
	Separators	41,627	120 scfd/separator	35,099
	Dehydrators	11,630	89.58 scfd dehydrator	7,324
	Meters/Piping	97,399	52.01 scfd/meter	35,609
South West	Associated Gas Wells	155,119	NA	NA
	Non-associated Gas Wells	13,860	37.24 scfd/well	3,628
	Gas Wells with Hydraulic Fracturing	27,627	37.24 scfd/well	7,232
	Heaters	11,243	58.97 scfd/heater	4,661
	Separators	23,316	125 scfd/separator	20,435
	Dehydrators	5,784	93.11 scfd dehydrator	3,786
	Meters/Piping	55,885	54.06 scfd/meter	21,237

Region	Activity	Activity Data	Emission Factor	Calculated Potential Emissions (MT)
West Coast	Associated Gas Wells	29,726	NA	NA
	Non-associated Gas Wells	1,999	42.49 scfd/well	597
	Gas Wells with Hydraulic Fracturing	95	42.49 scfd/well	28
	Heaters	2,094	67.29 scfd/heater	991
	Separators	1,529	142 scfd/separator	1,529
	Dehydrators	292	106 scfd dehydrator	218
	Meters/Piping	3,994	61.68 scfd/meter	1,732
Gulf Coast	Associated Gas Wells	39,709	NA	NA
	Non-associated Gas Wells	27,024	7.96 scfd/well	1,512
	Gas Wells with Hydraulic Fracturing	49,862	7.96 scfd/well	2,789
	Heaters	17,222	64.60 scfd/heater	7,821
	Separators	50,591	136.57 scfd/separator	48,571
	Dehydrators	10,719	102.00 scfd dehydrator	7,686
	Meters/Piping	90,288	59.21 scfd/meter	37,584

[a] Derived from ANNEX 3 Methodological Descriptions for Additional Source or Sink Categories (U.S. EPA, 2014).

The gas processing and gas transmission segments are not broken into regions like the gas production segment in the 2014 GHG Inventory. Instead, these segments provide national level emission estimates for their individual emission sources. For both segments, leak emissions include emissions from all components in the gas plants and on compression systems. The transmission segment leaks include leaks from transmission pipelines. The 2014 GHG Inventory calculates potential methane emissions from these sources using emission factors from the GRI/EPA study (GRI/U.S. EPA, 1996) and a 2010 ICF International (ICF) memo to the EPA on centrifugal compressors (ICF, 2010). The GHG Inventory emissions are calculated by applying the emission factors to activity counts (in this case, gas plants, compressor station counts, compressor counts, and pipeline miles) for each year of the inventory. Because some component counts are not available for each year of the GHG Inventory, a set of industry activity data drivers was developed and used to update activity data.[9] The 2014 GHG Inventory gas processing and gas transmission sources, emission factors, and methane emissions are presented in Table 2-7. For 2012, the 2014 GHG Inventory estimated 33,681 MT of potential methane emissions from gas processing leak emissions and 114,348 MT of potential methane emissions from gas transmission leak emissions.

Table 2-7. 2011 Data and Calculated Methane Potential Leak Emissions for the Natural Gas Processing and Natural Gas Transmissions Segments[a]

Segment	Activity	Activity Data	Emission Factor	Calculated Potential Emissions (MT)
Gas Processing	Plants	606	7,906 scfd/plant	33,681
Gas Transmission	Pipeline Leaks	303,126	1.55 scfd/mile	3,311
	Station	1,799	8,778 scfd/station	111,037

[a] Derived from ANNEX 3 Methodological Descriptions for Additional Source or Sink Categories, pg. A-177 (U.S. EPA, 2014).

For 2012, the 2014 GHG Inventory data estimates that potential emissions from leaks in production, processing and transmission are approximately 480,691 million MT of methane or about 8% of overall potential methane emissions from oil and gas.

2.5 Measurements of Methane Emissions at Natural Gas Production Sites in the United States (Allen et al., 2013)

A study completed by multiple academic institutions and consulting firms was conducted to gather methane emissions data at onshore natural gas sites in the U.S. This study used direct

---

[9] For example, individual compressor counts and compressor station counts are not available. Instead, these are obtained using a ratio of compressors to gas plants (for processing) and ratios of stations to pipeline miles and compressors to pipeline miles (for transmission) in the base year 1992. The 1992 ratios are then multiplied by the activity drivers, i.e., gas plant count or miles of pipeline, in the current year to estimate activity in current year.

measurements of methane emissions at 190 onshore natural gas sites in the U.S. (150 production sites, 27 well completion flowbacks, 9 well unloadings, and 4 workovers). The study covered the natural gas production segment.

For leak emissions, the study collected emissions data from 150 sites, 146 sites with wells and 4 sites with separators and other equipment on site. Leak emissions data from piping, valves, separators, wellheads, and connectors are provided in Table 2-8. The first step used to identify leaks from natural gas production sites was to scan the site using an OGI camera. The threshold for detection of a leak with the camera was 30 g/hr (Allen et al., 2013). After leaks were identified by the camera, the flow rate and the concentration of the leaks were measured using a Hi-Flow Sampler™ and the mass emission rate calculated. The instrument was calibrated using samples consisting of pure methane in ambient air. To account for the effect of ethane, propane, butane and higher alkanes on the leak measurements, gas composition data were collected for each natural gas production site that was visited. Based on the gas composition, the percentage of carbon accounted for by methane in the sample stream was determined. This percentage, multiplied by the total gas flow rate reported by the instrument, was the methane flow.

Table 2-8. Summary of Emissions from Leaks

	Emissions Per Well[a]				
	Appalachian	**Gulf Coast**	**Midcontinent**	**Rocky Mountain**	**All Facilities**
Number of Sites with Wells Visited (number of wells with leaks detected)	47 (30)	54 (31)	26 (19)	19 (17)	146 (97)
Methane Emission Rate (scf/min/well)	0.098 ± 0.059	0.052 ± 0.030	0.046 ± 0.024	0.035 ± 0.026	0.064 ± 0.023
Whole Gas Emissions Rate (based on site specific gas composition) (scf/min/well)	0.100 ± 0.060	0.058 ± 0.033	0.055 ± 0.034	0.047 ± 0.034	0.070 ± 0.024

[a] All leaks detected with the OGI camera, and does not include emissions from pneumatic pumps and controllers.

The study authors concluded the average values of leak emissions per well reported in Table 2-8 are comparable to the average values of potential emissions per well for gas wells, separators, heaters, piping and dehydrator leaks (0.072 scf methane/min/well) from the 2013 GHG Inventory, calculated by dividing the potential emissions in these categories in the 2013 GHG Inventory by the number of wells (Allen et al., 2013).

2.6   City of Fort Worth Natural Gas Air Quality Study (ERG, 2011)

The city of Fort Worth solicited a study that reviewed air quality issues associated with natural gas exploration and production. The goals of the study were to answer the following four questions:

- How much air pollution is being released by natural gas exploration in Fort Worth?

- Do sites comply with environmental regulations?

- How do releases from these sites affect off-site air pollution levels?

- Are the city's required setbacks for these sites adequate to protect public health?

To answer these questions, the study collected ambient air monitoring and direct leak and vented emissions measurements and performed air dispersion modeling. The study collected data from 375 well pads, 8 compressor stations, a gas processing plant, a saltwater treatment facility, a drilling operation, a hydraulic fracturing operation, and a completion operation. The point source test data was collected using an OGI camera, a toxic vapor analyzer (TVA), a Hi-Flow Sampler™ and stainless steel canisters. Each site was surveyed with an OGI camera and, if a leak was observed by the camera, the concentration of the leak was measured using the TVA. In addition, 10% of the total valves and connectors and the other components were surveyed using the TVA to determine leaks at or above 500 ppmv. The emission rates of the leaks identified by the OGI camera and the TVA survey were determined using a Hi-Flow Sampler™ to measure the volumetric flow rate of the leak. Gas samples from selected leaks were collected in stainless steel canisters for VOC and HAP analysis by a gas chromatograph/mass spectrometer (GC/MS).

Based on the results of the point source leak survey, the study estimated the total organic emissions to be 20,818 tons per year or 18,819 megagrams per year (Mg/yr), with well pads accounting for more than 75% of the total emissions. Hydrocarbons with low toxicities (methane, ethane, propane, and butane) accounted for approximately 98% of the emissions from this study. A summary of the average and maximum emissions from each of the site types is provided in Table 2-9. Table 2-10 provides a summary of the measured emissions by equipment type (e.g., connector, valve, other). Valves include manual valves, automatic actuation valves, and pressure relief valves. Connectors include flanges, threaded unions, tees, plugs, caps and open-ended lines where the plug or cap was missing. The category "Other" consists of all remaining components such as tank thief hatches, pneumatic valve controllers, instrumentation, regulators, gauges, and vents.

Table 2-9. Average and Maximum Point Source Emission Rates by Site Type[a]

Site Type	TOC (tons/yr)		VOC (Tons/yr)	
	Average	Maximum	Average	Maximum
Well Pad	16	445	0.07	8.6
Well Pad with Compressor(s)	68	4,433	2	22
Compressor Station	99	276	17	43
Processing Facility	1,293	1,293	80	80

a - Derived from Table 3.5-1 (ERG, 2011).

Table 2-10. Average and Maximum Point Source Emission Rates by Equipment Type[a]

Equipment Type	Methane (lb/yr)		VOC (lb/yr)	
	Average	Maximum	Average	Maximum
Connectors	8,918	169,626	27.6	171
Other	20,914	497,430	142	4,161
Valves	27,585	570,083	29.7	123

a - Derived from Emissions Calculation Workbook spreadsheet.

Some general observations of the well pad data provided in the Fort Worth report are:

- At least one leak was detected at 283 out of the 375 well pads monitored with an OGI technology with an average of 3.2 leaks detected per well pad;

- The TVA detected at least one leak greater than 500 ppm at 270 of the 375 well pads that were monitored with an average of 2.0 leaks detected per well pad;

- The number of wells located on well pads ranged from 0 to 13 with the average number of wells being 2.98 with a 99% confidence level of 0.31;

- The average number of components at each well site was 212 valves, 1596 connectors, 3 storage tanks, and 0.4 compressors;

- 124 out of the 375 well pads had at least one compressor onsite;

- There were 17 different owners of the 375 well sites in the Fort Worth area with the average number of well sites per owner being 22;

- Of the 1,330 leaks that were detected using either OGI technology or the TVA, 200 (15%) were classified as connector type leaks, 90 (7%) were classified as valve type leaks, and 1,040 (78%) were classified as other type leaks.

- Of these 1,330 leaks that were detected using OGI technology or the TVA, 1,018 (77%) were classified as non-tank leaks and the remaining 312 (23%) were classified as tank leaks.

2.7   Measurements of Well Pad Emissions in Greeley, CO (Modrak, 2012)

An onsite direct measurement study of emissions from 23 well pads in areas near

Greeley, CO (Weld County) was performed over a one-week period in July 2011. This study used the same source testing contractor and non-invasive leak detection and measurement procedures (OGI and Hi-Flow Sampler™) as in the City of Fort Worth Natural Gas Air Quality Study (ERG, 2011). Other than the number of production pads investigated (375 vs. 23), there were three major differences in the studies.

- The City of Forth Worth Air Quality Study was conducted in a predominately dry gas area of the Barnett shale whereas the Greeley study was conducted in an area with much higher relative condensate/oil production rates (wet gas). A typical leak or vented emission in a dry gas area is likely to have a higher methane to VOC ratio compared to an emission in a wet gas area.

- The State of Colorado requires emissions from condensate/oil tanks to be collected and controlled (e.g. routed to an enclosed combustors). In the City of Forth Worth Air Quality Study, most storage tanks contained produced water and were not controlled.

- The City of Fort Worth Air Quality Study used the EPA Compendium Method TO-15 and ASTM 1945 (for methane) for source canister analysis, whereas the Greeley study used the Ozone Precursor method (EPA/600-R-98/161) coupled with ASTM 1946/D1945 analysis of methane, ethane and propane. The canister analysis set used in the Greeley study had significantly more overlap for oil and gas product-related compounds (i.e. ethane, propane, other alkanes), whereas the TO-15 method provided more coverage for HAP compounds.

The objectives of the limited scope Greeley well pad study were to improve understanding of methane and speciated VOC emissions and investigate the use of commercially available non-invasive measurement approaches for application to wet gas production operations (including tank emissions).

The average production pad in the Greeley study consisted of 5 wells, 258 valves, 2,583 connectors, 3 condensate tanks, 1 produced water tank, 4 thief hatches, 5 pressure relief devices, 3 separators and 1 enclosed combustor control device. A total of 93 emission points were found with OGI technology at the 23 production sites and the emission rates were measured using a high volume sampler with a subset of 33 additionally sampled using evacuated canisters. A disproportionate number of detected emissions were found to be associated with storage tanks (72%). For the purposes of this white paper, a tank-related air emission is considered a leak if it exceeds the state or local emission limits. The study authors concluded condensate tank-related emissions observed in the Greeley study were not effectively collected and controlled. However, due to single point and instantaneous nature of the measurements, it is not known if these uncollected emissions exceed the state allowance.

Considering only emissions measurements with canister analysis, the average methane emissions from all storage tanks, excluding samples of known flash emissions, were much lower in the Greeley study compared to the City of Fort Worth Air Quality Study, 0.77 tons/year (n=21) and 21.9 tons/year (n=54), respectively. In contrast, the average VOC tank related emissions were much higher in the Greeley study compared to the City of Fort Worth Air Quality Study, 5.38 tons/year and 0.48 tons/year, respectively. Non-tank emissions followed similar trends: emissions of methane were higher in the City of Fort Worth Air Quality Study (7.73 tons/year (n=92) and 1.01 tons/year in the Greeley study (n=5)), while VOC emissions were higher in the Greeley study (0.46 tons/year in the Greeley study and 0.02 tons/year in the City of Fort Worth Air Quality Study). The authors noted that these emission estimates are based on instantaneous measurements. Because tank-related emissions vary diurnally and by season and may contain a residual flash emissions component, the extrapolation to yearly values (i.e., tons/year) is for informational purposes only and should not be used for comparison to permit or control limits. A journal article with additional analysis of these studies is in preparation (Modrak, 2012; Brantley et al., 2014a).

2.8   Quantifying Cost-Effectiveness of Systematic Leak Detection and Repair Programs Using Infrared Cameras (CL, 2013)

The study presented a summary of 4,293 surveys from two private sector firms that provide gas emission detection and measurement services to oil and gas facilities in the U.S. and Canada. These surveys only covered certain regions of the U.S. and Canada. The surveys included three categories of facilities: gas processing plants (614 surveys), compressor stations (1,915 surveys; includes both gas transmission and gas gathering systems), and well sites (1,764 surveys; includes single well heads and sites with up to 15 well heads). The surveys were conducted using OGI technology to locate leaking components and the leak rates were measured using a high-volume sampler. In some cases, where the facility owners did not need a precise volume measurement or where the leaking component was difficult to access for measurement, an estimate (evaluated visually using OGI technology based on the extensive experience of the operators) was used to make the decision to repair.

The study found that of the 58,421 components that were identified in the surveys, 39,505 (68%) were either leaking or venting gas. A summary of the leak rates for each of the categories is provided in Table 2-11. As the table shows, the study found that gas processing plants had the highest leak rate, followed by compressor stations and then well sites. The study noted that vents are the most common source of gas emissions from the identified emission sources, and about 40% of the vent emissions come from instrument controllers and compressor rod packing. Other vent sources come from production/storage tanks, lube oil vents, compressors, pumps, and engines. (Note: vented emissions are not considered leaks for the purposes of this paper).

Table 2-11. Distribution of Facilities Within Each Category by Leak Rate (in Mcf of gas per facility per year)[a]

Category	No leaks	≤ 99	100-499	500-1499	≥ 1500
Gas processing plants	3%	17%	32%	25%	23%
Compressor stations	11%	30%	36%	15%	9%
Well sites & well batteries	36%	38%	18%	5%	2%

a - Derived from Table 3 (CL, 2013).

The study results show that, for the facilities in the study, gas processing plants are the most likely to have leaks and the most likely to have large leaks, followed by compressor stations, and, lastly, well sites.

2.9   Mobile Measurement Studies in Colorado, Texas, and Wyoming (Thoma, 2012)

As will be described in detail in Section 3.4, emerging mobile measurement technologies are providing new capability for detection and measurement of emissions from upstream oil and gas production and other sectors. The EPA developed and applied one such mobile inspection technique as part of its Geospatial Measurement of Air Pollution (GMAP) program. (Thoma, 2012; Brantley et al., 2014b). Designed to be a rapidly-deployed inspection approach that can cover large areas, OTM 33A can locate unknown emissions (e.g., pipeline leaks or malfunctions) and can provide an emission rate assessment for upstream oil and gas sources, such as well pads located in relatively open areas. With measurements executed from stand-off observation distances of 20 m to 200 m, the mobile approach is not as accurate as onsite direct measurements but can provide source strength assessments with an accuracy of +/- 30% under favorable conditions with repeat measurements. OTM 33A relies on statistically representative downwind plume sampling, relatively obstruction-free line of sight observation, and a knowledge of the distance to the source (Thoma, 2012; Brantley et al., 2014b).

The EPA used OTM 33A to conduct several survey field campaigns in Weld County, CO in July 2010 and July 2011; areas near Fort Worth, TX (Wise, Parker, Tarrant, and Denton Counties) in September 2010 and 2011; in Sublette County, WY in June 2011, July 2012 and June 2013; and in the Eagle Ford, TX area (Maverick, Dimmit, La Salle, Webb, and Duval Counties) in September 2011. A total of 84 methane emission assessments were conducted in the Fort Worth area, 216 in WY, 93 in CO, and 22 in the Eagle Ford with offsite canister acquisition. Additionally, VOC emission estimates were executed at approximately 46% of these measurements. A subset of these field studies are described in (Thoma, 2012) with an expanded discussion, and slight revision of results to be published in (Brantley et al., 2014b). These data are primarily from well pads and represent an integration of all emissions (leak and vented) on the site. (Note: Vented emissions are not defined as "leaks" in this paper, therefore, the emission rates presented below include emissions that are not considered leaks in this paper). The study authors note, as with all instantaneous measurement approaches, the OTM 33A assessment may capture emissions that are short-term in nature (i.e., flash emissions) so extrapolation to annual emissions is difficult.

The preliminary results from the study (Thoma, 2012) show median methane emission rates of 0.21 grams per second (g/s), 0.43 g/s and 0.79 g/s and VOC emission rates of 0.16 g/s, 0.04 g/s and 0.30 g/s for the CO, TX, and WY studies, respectively (excluding Eagle Ford).

The study authors note that using improved analysis procedures, the above median rates will likely be revised slightly lower in a future publication. Offsite OGI was used in many cases to positively identify the origin of emissions. The study authors concluded that many of the high emission values were attributed to maintenance-related issues such as open thief hatches, failed pressure relief valves, or stuck dump valves. The difference in VOC emissions between the TX studies and the CO and WY studies is a result of the natural gas from the TX well sites being a dry natural gas. Additional analysis of the emission measurements including comparisons to natural gas, condensate/oil, and produced water production will be contained in a forthcoming article (Brantley et al., 2014b).

2.10   Economic Analysis of Methane Emission Reduction Opportunities in the U.S. Onshore Oil and Natural Gas Industries (ICF International, 2014)

The Environmental Defense Fund (EDF) commissioned ICF to conduct an economic analysis of methane emission reduction opportunities from the oil and natural gas industry to identify the most cost-effective approach to reduce methane emissions from the industry. The study projects the estimated growth of methane emissions through 2018 and focuses its analysis on 22 methane emission sources in the oil and natural gas industry (referred to as the targeted emission sources). These targeted emission sources represent 80% of their projected 2018 methane emissions from onshore oil and gas industry sources. Well site leaks (includes heaters, separators, dehydrators and meters/piping) and pipeline leaks are two of the 22 emission sources that are included in the study.

The study relied on the 2013 GHG Inventory for methane emissions data for the oil and natural gas sector. The emissions data were revised to include updated information from the GHGRP (U.S. EPA, 2013) and the *Measurements of Methane Emissions at Natural Gas Production Sites in the United States* study (Allen et al., 2013). The revised 2011 baseline methane emissions estimate was used as the basis for projecting onshore methane emissions to 2018. One of the major differences in the revised 2011 baseline methane emissions estimate developed by ICF is the inclusion of a separate category for gathering and boosting operations. The 2013 GHG Inventory includes gathering and boosting operations in the onshore production segment and is based on the GRI/EPA measurement study (GRI/U.S. EPA, 1996).

The 2011 baseline methane inventory developed by ICF used the wellhead emission factor developed from the University of Texas study (Allen et al., 2013) to estimate leak emissions from well sites, which was reported as 97.6 scf/day. This emissions factor was applied to the natural gas well counts obtained from World Oil magazine to estimate the total methane leak emissions from well sites. These changes resulted in an estimated 14 billion cubic feet (264,000 MT) of methane emissions from wellheads in comparison.

Leak emissions from heater, separators, dehydrators, and meters/piping in the natural gas production sector were calculated using the GRI/EPA emissions factors for each of these emission sources. The study estimated methane emissions were 15 billion cubic feet (283,000 MT) from these sources.

Natural gas processing plant leak emission were determined by ICF using data from the GHGRP (U.S. EPA, 2013) and a list of processing plants maintained by the EIA. The study by ICF determined that there are 909 gas processing and treatment facilities in the U.S. The study estimated methane emissions from processing facilities to be 3 billion cubic feet (56,600 MT).

The study did not provide specific equipment leak information for the natural gas transmission and storage sectors. However, the report did provide information on pipeline leaks from transmission of natural gas. The report estimated methane emissions of 0.2 billion cubic feet (3,800 MT).

The estimate of total national emissions from leaks in the natural gas production, processing, transmission and storage segments for 2011 was 604,000 MT of methane.

2.11   Identification and Evaluation of Opportunities to Reduce Methane Losses at Four Gas Processing Plants (Clearstone, 2002)

This study, referred to as "Clearstone I,"[10] presented the results of the implementation of a comprehensive directed inspection and maintenance (DI&M) program at four gas processing

---

[10] "Identification and Evaluation of Opportunities to Reduce Methane Losses at Four Gas Processing Plants." Prepared for GTI and the U.S. EPA under grant 827754-01-0, by Clearstone Engineering. June 20, 2002. Also, note that a follow-up study, referred to as Clearstone II, was released in 2006, which studied five processing plants, one being a repeat from the plants studied in Clearstone I.

plants in the western U.S. in 2000. The work done during this study involved a survey of all gas service equipment components, as well as the measurement or engineering calculation of gas flows into the vent and flare systems. This study did not focus on hydrocarbon liquid services. In total, 101,193 individual gas service components were screened, along with 5 process vents, 28 engines, 7 process heaters, and 6 flare/vent systems.

The leak survey was conducted using bubble tests with soap solution, portable hydrocarbon gas detectors, and ultrasonic leak detectors. A screening value of 10,000 ppm or greater was used as the leak definition. The majority of components were screened using soap solution, but if a component was determined to be emitting gas, a hydrocarbon gas analyzer was used to determine if the component would be classified as a leaker per the above definition. Most leak rates were measured using a Hi-Flow™ Sampler, unless the leak was above the upper limit of the unit's design (14 m^3/hour). If the Hi-Flow™ Sampler could not be used, bagging or other direct measurement techniques were used, as appropriate.

From the survey, approximately 2,630 of the 101,193 screened components (2.6%) were determined to be leaking. The study states that "components in vibrational, high-use or heat-cycle gas service were the most leak prone." The majority of the leaks were attributed to a relatively small number of leaking components. Table 2-12 presents the breakdown of leak emissions by component type.

Table 2-12. Distribution of Natural Gas Emissions from Leaking Component Types

Component Type	Percent of Leak Emissions
Valves	30.0%
Connectors	24.4%
Compressor Seals[a]	23.4%
Open-Ended Lines	11.1%
Crankcase Vents (on Compressors)	4.2%
Pressure Relief Valves	3.5%
Other (Pump Seals, Meters, Regulators)	3.4%

The study also provided an analysis of the payback periods for fixing the identified leaks. That analysis is discussed in Section 3.2 of this paper.

2.12 Cost-Effective Directed Inspection and Maintenance Control Opportunities at Five Gas Processing Plants and Upstream Gathering Compressor Stations and Well Sites (Clearstone, 2006)

This study, referred to as "Clearstone II,"[11] presented the results of a comprehensive emissions measurement program at 5 gas processing plants, 12 well sites, and 7 gathering stations in the U.S. in 2004 and 2005. This work was done as follow up on a study done in 2000, referred to as Clearstone I, in which four gas processing plants were surveyed. (Note: one of the gas processing plants surveyed in the Clearstone I study was also surveyed in the Clearstone II study.) The work done involved a survey of all gas service equipment components at these 24 sites. The goal was to identify cost-effective opportunities for reducing natural gas losses and process inefficiencies. In total, 74,438 individual components were screened.

The leak survey was conducted using bubble tests with soap solution, portable hydrocarbon gas detectors, and ultrasonic leak detectors. A screening value of 10,000 ppm or greater was used as the leak definition. The majority of components were screened using soap solution, but if a component was determined to be emitting gas, a hydrocarbon gas analyzer was used to determine if the component would be classified as a leaker per the above definition. Most leak rates were measured using a Hi-Flow™ Sampler, unless the leak was above the upper limit of the unit's design (14 m^3/hour). For consistency, both the Clearstone I and Clearstone II surveys used the same Hi-Flow™ Sampler. If the Hi-Flow™ Sampler could not be used, bagging or other direct measurement techniques were used, as appropriate.

_____

[11] "Cost-Effective Directed Inspection and Maintenance Control Opportunities at Five Gas Processing Plants and Upstream Gathering Compressor Stations and Well Sites." Prepared for the U.S. EPA under grant XA-83046001-1, by National Gas Machinery Laboratory, Clearstone Engineering, and Innovative Environmental Solutions, Inc. March 2006. Note: This study, referred to as "Clearstone II", was a follow up to a study released in 2002, referred to as "Clearstone I," which surveyed four processing plants, one of which was resurveyed in Clearstone II.

34

Secondarily to the above leak detection methodology, for all five surveys of gas processing plants in the study, OGI cameras were also used in order to compare the performance of the OGI cameras with conventional leak detection methods. Although no quantitative comparison was done, the study concluded that the cameras are able to screen components about three times as quickly as the other methods, find leaks that are inaccessible to the other methods, and allow for rapid leak source identification.

From the survey, approximately 1,629 of the 74,438 screened components (2.2%) were determined to be leaking. The study states, similarly to Clearstone I, that "components in vibrational, high-use, and heat-cycle gas service were the most leak prone." Further, the majority of the leak emissions could be attributed to a relatively small number of the leaking components. Table 2-13 presents the breakdown of natural gas leak emissions by component type.

Table 2-13. Distribution of Natural Gas Emissions from Leaking Component Types

Component Type	Percent of Leak Emissions
Open-Ended Lines	32%
Connectors	30%
Compressor Seals	20%
Block Valves	15%
Other (PRVs, Meters, Regulators, etc.)	3%

The study also provides a comparison for the one gas plant that was surveyed in both studies. This plant was resurveyed in order to investigate changes in its leak characteristics. It was noted that about 30% of the equipment components in the plant had been decommissioned between the surveys due to the replacement of old process units with newer ones. Generally, the replacement process units and equipment components had substantially reduced emission rates compared to the decommissioned units. The overall reduction for the new units was an 80% decrease in total hydrocarbon (THC) emissions compared to the old units. However, the THC emissions for the plant as a whole increased about 50% between the two surveys. The study gives several possible reasons for this, including the fact that the five-year timeframe between surveys exceeded the mean repair life for most of the components. The study also states that there may have been inadequate follow-up to maintenance recommendations provided during the first survey, as the documentation of repairs indicated it was "unclear what maintenance activities were undertaken in response to the Phase I survey."

3.0   AVAILABLE EMISSIONS MITIGATION TECHNIQUES

There are a number of technologies available that can be used to identify leaks and a number of approaches to repairing those leaks. The technologies for identifying leaks and the approaches to repairing leaks are discussed in separate sections below.

3.1   Leak Detection

A variety of approaches are used for leak detection. For many regulations with leak detection provisions, the primary method for monitoring to detect leaking components is EPA Reference Method 21 (40 CFR Part 60, Appendix A). Method 21 is a procedure used to detect VOC leaks from process equipment using an analyzer, such as a TVA or an OVA. In addition, other monitoring tools such as OGI cameras, soap solution, acoustic leak detection, ambient monitors and electronic screening devices can be used to monitor process components. A summary of these technologies is presented below.

3.1.1   Portable Analyzers

Description

A portable monitoring instrument is used to detect hydrocarbon leaks from individual pieces of equipment. These instruments are intended to locate and classify leaks based on the leak definition of the equipment as specified in a specific regulation, and are not used as a direct measure of mass emission rate from individual sources. The instruments provide a reading of the concentration of the leak in either ppm, parts per billion (ppb), or percent concentration. For portable analyzers, EPA Reference Method 21 requires the analyzer to respond to the compounds being processed, be capable of measuring the leak definition concentration specified in the regulation, be readable to ±2.5% of the specified leak definition concentration and be equipped with an electrically driven pump to ensure that a sample is provided to the detector at a constant flow rate.

The portable analyzers can be used to estimate the mass emissions leak rate by converting the screening concentration in ppm to a mass emissions rate by using the EPA correlation equations from the Protocol for Equipment Leak Emission Estimates (U.S. EPA, 1995). The correlation equations in the Protocol can be used to estimate emissions rates for the entire range of screening concentrations, from the detection limit of the instrument to the "pegged" screening concentration, which represents the upper limit of the portable analyzers (U.S. EPA, 2003a).

The portable analyzers must be calibrated using a reference gas containing a known compound at a known concentration. Methane in air is a frequently used reference compound. The calibration process also determines a response factor for the instrument, which is used to correct the observed screening concentration to match the actual concentration of the leaking compound. For example, a response factor of "one" means that the screening concentration read by the portable analyzer equals the actual concentration at the leak (U.S. EPA, 2003a). Screening concentrations detected for individual components are corrected using the response factor (if necessary) and are entered into the EPA correlation equations to extrapolate a leak rate measurement for the component (U.S. EPA, 2003a).

Applications

The portable monitoring instruments operate on a variety of detection principles, with the three most common being ionization, IR absorption and combustion (U.S. EPA, 1995). The ionization detectors operate by ionizing the sample and then measuring the charge (i.e., number of ions) produced. Two methods of ionization currently used are flame ionization and photoionization. A standard flame ionization detector (FID) measures the total carbon content of the organic vapor sampled. Certain portable FID instruments are equipped with gas chromatograph (GC) options making them capable of measuring total gaseous non-methane organics or individual organic components (U.S. EPA, 1995). The photoionization detector (PID) uses ultraviolet light (instead of a flame) to ionize organic vapors. As with FIDs, the detector response varies with the functional group in the organic compounds. Photoionization detectors have been used to detect leaks in process units in the Synthetic Organic Chemical Manufacturing Industry (SOCMI), especially for certain compounds, such as formaldehyde, aldehydes, and other oxygenated compounds, which may not give a satisfactory response on a FID or combustion-type detector (U.S. EPA, 1995).

Nondispersive infrared (NDIR) instruments operate on the principle of light absorption characteristics of certain gases. These instruments are usually subject to interference because other gases, such as water vapor and CO_2, may also absorb light at the same wavelength as the compound of interest (U.S. EPA, 1995). These detectors are generally used only for the detection and measurement of single components. For this type of detection, the wavelength at which a certain compound absorbs IR radiation is predetermined and the device is preset for that specific wavelength through the use of optical filters (U.S. EPA, 1995).

Combustion analyzers are designed either to measure the thermal conductivity of a gas or to measure the heat produced by combustion of the gas. The most common method in which portable VOC detection devices are used involves the measurement of the heat of combustion. These detection devices are referred to as hot wire detectors or catalytic oxidizers. Combustion analyzers, like most other detectors, are nonspecific for gas mixtures (U.S. EPA, 1995). In addition, combustion analyzers exhibit reduced response (and, in some cases, no response) to gases that are not readily combusted, such as formaldehyde and carbon tetrachloride (U.S. EPA, 1995).

The typical types of portable analyzers used for detecting leaks from components are OVAs and TVAs. An OVA is an FID, which measures the concentration of organic vapors over a range of 9 to 10,000 ppm (U.S. EPA, 2003a). A TVA combines both a FID and a PID and can measure organic vapors at concentrations exceeding 10,000 ppm. Toxic vapor analyzers and OVAs measure the concentration of methane in the area around a leak (U.S. EPA, 2003a).

Screening is accomplished by placing a probe inlet at an opening where leakage can occur. Concentration measurements are observed as the probe is slowly moved along the interface or opening, until a maximum concentration reading is obtained. The maximum concentration is recorded as the leak screening value. Screening with TVAs and OVAs can be a slow process, requiring approximately one hour for every 40 components, and the instruments require frequent calibration.

Costs

The costs of the portable analyzers vary based on the type of analyzer used to measure leak concentrations. The documentation for the EPA National Uniform Emission Standards for Equipment Leaks (40 CFR part 65, subpart J) provides a cost of $10,800 for a portable monitoring analyzer (RTI, 2011). Additional costs would also include labor costs associated with performing the screening and would depend on the number of components screened.

3.1.2   Optical Gas Imaging (IR Camera)

Description

Optical gas imaging (OGI) is a technology that operates much like a consumer video-camcorder and provides a real-time visual image of gas emissions or leaks to the atmosphere. The OGI camera works by using spectral wavelength filtering and an array of IR detectors to visualize the IR absorption of hydrocarbons and other gaseous compounds. As the gas absorbs radiant energy at the same waveband that the filter transmits to the detector, the gas and motion of the gas is imaged. The OGI instrument can be used for monitoring a large array of equipment and components at a facility, and is an effective means of detecting leaks when the technology is used appropriately. The EPA has worked extensively with OGI technology and is in the process of further evaluating its capabilities. Information presented below, unless otherwise cited, is based on that evaluation work.

Applications

The detection capability of the OGI camera is based on a variety of factors such as detector capability, gas characteristics of the leak, optical depth of the plume and temperature differential between the gas and background. The EPA is currently studying OGI technology in order to determine its capabilities and limitations.

The OGI system provides a technology that can potentially reduce the time, labor and costs of monitoring components. The capital cost of purchasing an OGI system is estimated to be $85,000 (Meister, 2009). The ICF economic analysis estimated the capital cost of the OGI system to be $124,000 (ICF International, 2014). The EPA estimated that the OGI can monitor 1,875 pieces of equipment per hour at a petroleum refinery (RTI, 2012). This study assumes for every hour of video footage, the operator would spend an additional 1.4 hours conducting activities for calibration, OGI adjustments, tagging leaks and other activities. Another estimate, (ICF Consulting, 2003) stated that OGI can monitor 35 components per minute (2,100 components per hour). In comparison, the average screening rate using a handheld TVA or OVA is roughly 700 components per day (ICF Consulting, 2003). However, the EPA's recent work with OGI systems suggests these studies underestimate the amount of time necessary to thoroughly monitor components for leaks using OGI technology. Additionally, the number of pieces of equipment that could be monitored per hour at an upstream oil and gas facility would likely be less than at a refinery given that equipment tends to be farther apart at these facilities than at a refinery.

By increasing the number of pieces of equipment that can be viewed per hour, the OGI system could potentially reduce the cost of identifying leaks in upstream oil and gas facilities when compared to using a handheld TVA or OVA. A recent study (CL, 2013) analyzed 4,293 leak detection surveys completed for the oil and gas industry using OGI systems. These surveys were completed by external contractors hired by the owner or operator of the oil and gas facility. This study estimated the average abatement cost to be approximately $0 per ton of VOC and approximately -$375 per ton of VOC for well sites and compressor stations, respectively. These estimates assume all leaks that are found are repaired and the recovered methane can be sold for $4/Mcf. The average costs of performing the OGI surveys in the study are $2,300 for a compressor station, $1,200 for multi-well batteries, $600 for single well batteries and $400 for well sites (CL, 2013). (Note: Only a prepublication draft was available of this report when the EPA was completing this white paper.)

Another advantage of OGI for detecting leaks is finding leaks not directly related to components while in the process of surveying the overall site. Leaks such as degradation in the exterior of tanks or leaks in lines buried underground would be seen with OGI but very hard to locate with a handheld TVA or OVA.

For the application of this technology to this sector, the gas characteristics are well suited for the typical OGI camera technology because the leaks tend to be almost all methane, alkane or aromatics. Methane, alkanes and aromatics are all detectable due to having carbon-hydrogen bonds.

OGI Operational Considerations

While the operator or inspector using OGI technology can see leaking emissions from equipment, quantifying the emissions is difficult. To quantify emissions with an OGI camera, extensive metadata, such as apparent background temperature, gas leak temperature, leak size and wind speed must also be taken. These parameters would then be used with a developed and evaluated algorithm to quantify emissions. The EPA is not aware of the existence or evaluation of such an algorithm at this time. However, in addition to algorithms, operators can use quantification equipment such as a Hi-Flow™ Sampler.

The OGI system is also sensitive to the ambient conditions around the equipment that is being inspected. The larger the temperature differential between the leaking gas and the contrasting background (e.g., sky, ground or equipment), the easier the leaking gas is to see. The apparent temperature of the sky, a commonly used background, is also highly dependent on weather conditions such as cloud cover, ambient temperature and relative humidity. Additionally, high or variable wind conditions can reduce the optical depth and make it difficult for gas leaks to be identified, because the gas plume is quickly carried away from the source of the leak. Both these characteristics could result in operators being unable to identify leaks if the ambient conditions are not optimal.

Lastly, the effectiveness of an OGI instrument is dependent on the training and expertise of the operator. Well-trained and experienced operators are able to detect leaks with the OGI system that lesser experienced operators do not detect.

Current OGI Usage in the Oil and Gas Industry

The EPA is not aware of any studies that estimate the extent of the usage of OGI systems in the oil and natural gas production sector. However, certain proposed and existing regulations allow OGI systems as an option for fulfilling leak detection requirements, and some companies are using the technology voluntarily such as through the Natural Gas STAR program. Additionally, the GHGRP subpart W allows for the use of OGI technology in some circumstances and the Alternative Work Practice regulation (40 CFR Part 60, subpart A) allow the use of OGI technology along with an annual Method 21 survey as an alternative to a traditional leak detection and repair (LDAR) program using Method 21.

The State of Colorado recently proposed regulations that would require leak inspections at all well sites, compressor stations upstream of the processing plant and storage vessels. These proposed regulations allow OGI inspections, Method 21 or other "[d]ivision approved instrument based monitoring device or method" to detect leaks (CO Department of Public Health and Environment, Air Quality Control Commission, Regulation Number 7, Proposed November 18, 2013).

The State of Wyoming, as part of its permitting guidance, requires facilities with emissions greater than 4 tpy of VOCs in the Upper Green River Basin, the Jonah-Pinedale Anticline Development Area and Normally Pressured Lance to conduct quarterly leak emissions inspections, and OGI inspections are allowed in addition to Method 21 inspections or audio-visual-olfactory inspections (Wyoming Department of Environmental Quality, Oil and Gas Production Facilities, Chapter 6, Section 2 Permitting Guidance, September 2013).

The Alberta Energy Regulator requires that a "licensee of a facility must develop and implement a program to detect and repair leaks." These programs must "meet or exceed" the Canadian Association of Petroleum Producer's (CAPP) best management practice (BMP) for leak emissions management (CAPP, 2011). The CAPP BMP allows OGI technology for performing these leak inspections (CAPP, 2007).

Lastly, the EPA has found that owners and operators are voluntarily using OGI systems to detect leaks. However, the EPA does not know the extent of these voluntary efforts within the industry on a national level.

3.1.3   Acoustic Leak Detector

Description

Acoustic leak detectors are used to detect the acoustic signal that results when pressurized gas leaks from a component. This acoustic signal occurs due to turbulent flow when pressurized gas moves from a high-pressure to a low-pressure environment across a leak opening (U.S. EPA, 2003a). The acoustic signal is detected by the analyzer, which provides an intensity reading on the meter. Acoustic detectors do not measure leak rates, but do provide a relative indication of leak size measured by the intensity of the signal (or how loud the sound is) (U.S. EPA, 2003a).

Applications

Generally, two types of acoustic leak detection methods are used; high frequency acoustic leak detection and ultrasound leak detection. High frequency acoustic detection is best applied in noisy environments where the leaking components are accessible to a handheld sensor (U.S. EPA, 2003a). Ultrasound leak detection is an acoustic screening method that detects airborne ultrasonic signals in the frequency range of 20 kHz to 100 kHz and can be aimed at a potential leak source from a distance of up to 100 feet (U.S. EPA, 2003a). Ultrasound detectors can be sensitive to background noise, although most detectors typically provide frequency tuning capabilities so that the probe can be tuned to a specific leak in a noisy environment (U.S. EPA, 2003a).

A URS Corporation/University of Texas at Austin (URS/UT) study described a "through-valve acoustic leak detection device" or VPAC that was used to measure leaks at six sites (four gathering/boosting stations and two natural gas processing plants) (URS/UT, 2011). Leak measurements were made using the VPAC device and high volume sample to compare the readings from the two devices. The study authors found that there was no statistically significant correlation between the VPAC and the direct flow measurements, and the study authors determined that the VPAC method was not considered to be an accurate alternative to direct measurement for the sources tested (URS/UT, 2011).

Costs

No cost data for acoustic leak detectors were available in the studies or research documents.

3.1.4   Ambient/Mobile Monitoring

Description

A growing number of research and industry groups are using mobile measurement approaches to investigate a variety of source emissions and air quality topics. For oil and natural gas applications, a vehicle can be equipped with at minimum a methane measurement instrument and GPS to facilitate discovery of previously unknown sources and in more advanced forms, provide information on source emission rates.

Applications

Mobile leak detection techniques sample emission plumes from stand-off (sometimes offsite) observing locations and are, therefore, generally less accurate than direct (onsite) source measurements. Mobile leak detection techniques can cover large survey areas and can be particularly useful in identifying anomalous operating conditions (e.g., pipeline leaks and well pad malfunctions) in support of onsite OGI and safety programs. All mobile techniques require downwind vehicle access and favorable wind conditions for plume transport to the observing location. The presence of trees or other obstructions can limit the efficacy of mobile leak detection techniques and in some cases prevent the application of remote source emission rate assessment.

Mobile leak detection instrument packages require some expertise for operation, especially in source emission rate measurement applications. Additionally, while mobile leak detection techniques can detect emissions around a site, such as a well site or gathering station, it cannot necessarily pinpoint the equipment that is the source of those emissions. Mobile leak detection techniques might be best used in conjunction with OGI technology; an OGI inspection would be triggered by the detection of above normal emissions by the mobile leak detection technique. In conversations with operators of upstream oil and natural gas facilities, the EPA has discovered that some companies are voluntarily using this two-phase approach to detect and then pinpoint VOC and methane leaks. It is believed that future forms of mobile leak detection techniques for the oil and gas sector may include lower cost, work truck-mounted systems that provide fully autonomous detection capability for anomalous emissions in support of such an onsite OGI inspection (Thoma, 2012).

An example of a mobile leak detection technique applicable to the upstream oil and gas sector is being developed under the EPA's Geospatial Measurement of Air Pollution (GMAP) program (Thoma, 2012). The near-field OTM 33A produces a 20-minute "snapshot" measure of emissions from near ground level point sources at observation distances of approximately 20 to 200 m. With strict application and favorable conditions, this type of point sensor-based remote measurement has source emission rate measurement accuracies in the $\pm$ 30% range with ensemble averages achieving accuracies within $\pm$ 15% by reducing random error effects. Although future, fixed deployment, low cost sensor systems may provide long-term emission level monitoring capability for oil and gas production sites,[12] current mobile assessment approaches can only provide a "snapshot" of emissions. Because some oil and gas upstream sources possess significant temporal and seasonal variability, the short-term nature of observation must be considered to avoid error in exportation of instantaneous emissions (e.g., to tons per year estimates). Results of well pad measurements from multiple oil and gas fields using mobile measurement are presented in Section 2.

Costs

Current mobile measurement instrument packages can range in cost from approximately $20,000 - $100,000 depending on the capability of the package.

3.2   Repair

After a leak is detected, the owner or operator of the facility must decide whether or not to fix the leak, unless they are required to fix the leak due to regulatory or permitting obligations.

---

[12] A collaborative request for proposal (RFP) was released in the spring of 2014 by Apache Corporation, BG Group, EDF, Hess Corporation, Noble Energy, and Southwestern Energy called the "Methane Detectors Challenge: Continuous Methane Leak Detection for the Oil and Gas Industry." The "Challenge" is "designed to spur the development of cutting-edge, new technologies that provide continuous detection of methane emissions." Available at: http://www.edf.org/energy/natural-gas-policy/methane-detectors-challenge

This decision can be based on several factors, including, the cost of fixing the leak and the size of the leak. A number of studies discuss costs and effectiveness of various leak repair options.

3.2.1   Quantifying Cost-Effectiveness of Systematic Leak Detection and Repair Programs Using Infrared Cameras (CL, 2013)

This study, discussed previously in Section 2, provided an analysis of the net present values (NPVs) of repairing all of the identified leaks in the surveys using the estimated repair cost and the value of the recovered gas. The study found that over 90% of gas emissions from leaks can be repaired with a payback period of less than one year, assuming a value of $3 per thousand cubic feet ($/Mcf) for the recovered gas. However, when compared with the cost of the monitoring (estimated to be $600 to $1,800 per facility), the economic benefits of repairing the leaks at most facilities are less than the total cost of the survey. For well sites and well batteries, the study estimated that 1,424 of the sites (81%) had a negative NPV, which averaged -$1,160 per facility. However, when the all of the individual well sites and well batteries are aggregated into a group, the aggregated NPV is positive, which suggests that a minority of sites have high leak rates and, thus, a positive NPV for monitoring and fixing leaks. These sites skew the mean NPV to a positive value.

The study also analyzed two alternative repair strategies: only repair leaks that are economic to repair (e.g., NPV > 0 for the repair) or repair of leaks that exceeded a certain threshold (e.g., 20 thousand cubic feet per year (Mcf/yr)). A summary of the findings for each of the scenarios is provided in Table 3-1.

Table 3-1. Comparison of Three Hypothetical Repair Strategies for Multi-Well Batteries[a]

Category	Repair all leaks	Repair leaks with a NPV>0	Repair leaks > 20 Mcf
Potential leak reductions after survey	94.5%	92.6%	88.1%
Methane abatement cost ($/ton CO_2e)	1	0.8	1.7
VOC abatement cost ($/ton VOC)	46	41	79
Average number of leaks to repair	3.8	3.5	2.9

a - Derived from Table 3 (CL, 2013).

The study concludes that the potential leak reductions after survey, methane abatement cost, VOC abatement cost, and average number of leaks to repair are similar under each of the three strategies. The study authors conclude that the results are similar because once a leak is found it is almost always economic to repair it.

The study also provided costs of repair and leak detection based on the survey data. The average cost of hiring an external service provider to perform a survey using OGI technology was determined to be $1,200 for multi-well batteries, $600 for single well batteries, and $400 for a well site. The range of costs of repair for well sites is shown in Table 3-2.

Table 3-2. Total Average Leak Rate and Repair Costs by Components at Well Sites

Component	Leak rate (cfm)	Repair Costs			
		Minimum	Average	Median	Maximum
Connector/Connection	0.11	$15	$56	$50	$5,000
Instrument Controller	0.03	$20	$129	$50	$2,000
Valve	0.04	$20	$90	$50	$5,500
Open-Ended Line	0.02	---[b]	---[b]	---[b]	---[b]
Regulator	0.02	$20	$189	$125	$1,000

a - Derived from Tables 6 and 7 (CL, 2013).
b – Repair costs for open-ended lines were not provided in the document.

3.2.2   Identification and Evaluation of Opportunities to Reduce Methane Losses at Four Gas Processing Plants (Clearstone, 2002)

The Clearstone I study, discussed in Section 2, provided analysis of the payback periods for fixing the identified leaks, and what level of emission reductions could be achieved. Overall, the study estimated that up to 95% of total natural gas losses can be reduced cost-effectively (assumed gas price of $4.50 per Mcf), which corresponds to methane reductions of nearly 80%. The study also presents scenarios where only those reduction opportunities having a certain payback period (e.g., 6 months or 1 year) are implemented. For those cases, the estimated percent of total natural gas loss reduction and corresponding reductions in methane are presented in Table 3-3. One caveat from the study is that the payback periods do not take into account the cost of the leak detection survey, only factoring in cost of repair and benefit of the gas captured.

Table 3-3. Achievable Emission Reduction Percentages for Given Positive Payback Periods

Emission Type Reduction	Payback Period			
	< 6 months	< 1 year	< 2 years	< 4 years
Natural Gas	78.8%	92.3%	93.1%	94.9%
Methane	71.9%	78.1%	79.2%	79.5%

The study estimated that implementing all of the cost-effective repair opportunities identified would result in gross annual cost savings of approximately $1.1 million across the plants in the study (based on a gas value of $4.50 per Mcf). This amounts to over 50% of total cost-effective loss reduction opportunities identified for all emission sources (leaks, flaring, combustion equipment, and storage tanks) at the plants, and results in an average annual net savings of approximately $280,000 per site (the site-specific values range between $180,000 and $330,000).

3.2.3 Cost-Effective Directed Inspection and Maintenance Control Opportunities at Five Gas Processing Plants and Upstream Gathering Compressor Stations and Well Sites (Clearstone, 2006)

The Clearstone II study, discussed in Section 2, analyzed the cost-effectiveness of repairing the leaks identified in the surveys that were performed. The study estimated that up to 96.6% of total natural gas losses could be reduced cost-effectively (assuming a gas price of $7.15 per Mcf), which corresponds to methane reductions of 61%. The study also estimated that the average annual lost gas values from the sites surveyed were $536,270 per gas plant, $49,018 per gathering station, and $3,183 per well site.

This study also provided the base repair cost and mean repair life for 16 types of components. The values for several of the more common components reported in the study are summarized in Table 3-4.

Table 3-4. Basic Repair Costs and Mean Repair Life for Several Common Leaking Components

Component Type	Basic Repair Costs		Mean Repair Life (years)
	Low	High	
Compressor Seals[a]	$2,000	$2,000	1
Flanges	$25	$400	2
Open-End Lines	$60	$1,670	2
Pressure Relief Valves	$79	$725	2
Threaded Connections	$10	$300	2
Tubing Connections	$15	$25	4
Valves	$60	$2,229	2 - 4
Vents	$2,000	$5,000	1

[a] For the purposes of this paper, compressor seal emissions are not considered leaks.

3.2.4 Natural Gas STAR Directed Inspection and Maintenance (U.S. EPA, 2003a, U.S. EPA, 2003b, and U.S. EPA, 2003c)

For detecting and repairing leaks, the Natural Gas STAR program recommends implementation of a DI&M program to economically reduce methane emissions from leaking components (U.S. EPA, 2003a, U.S. EPA, 2003b, and U.S. EPA, 2003c). A DI&M program, which can be implemented at any facility in the upstream or downstream sector of the industry, starts with a comprehensive baseline emissions survey. This survey involves screening all of the components at the facility to identify the leaking components, as well as measuring the identified leaks to determine emission rates. Determining an emissions rate is an important step that allows the economic evaluation of mitigation techniques. Natural Gas STAR partners have reported using OGI technology to effectively scan large numbers of components in a short span of time. The choice of leak detection equipment typically depends on the number of components to be scanned. Optical gas imaging technology is popular at facilities that have thousands of components, such as at processing plants. From previous field studies conducted by the EPA and Natural Gas STAR partners, the EPA has observed that typically 20% of the top leaking components account for approximately 80% of the emissions from a facility. This provides a strong basis to conduct DI&M at facilities because fixing a small number of leaks can significantly reduce the total leak emissions from a facility.

Once the leaking sources have been identified, the next step recommended is the economic analysis of mitigation techniques. The estimated repair costs for the identified leaks can be compared to the potential savings from fixing the leaks based on the value of natural gas, and the leaks that are determined to be economical to fix by the owner can be repaired.

Not all leaks identified can be fixed immediately. For example, leaks on a flange on a transmission pipeline cannot be fixed without shutting down the system and purging the pipeline of all the natural gas. The identification of leaks before a shutdown through a DI&M program helps facilities focus on specific areas during a shutdown cycle. Shutdown cycles are usually short, lasting from a day up to a week.

The Natural Gas STAR program also lists average emission rates, repair cost ranges, and payback periods for fixing leaks at several different facilities. Tables 3-5 and 3-6 show the emission rates and repair costs for several common leaking components at gas processing plants, transmission compressor stations, and gate stations.

Table 3-5. Total Average Leak Rate and Repair Costs by Component at Processing Plants

Component	Average Component Leak Rate by Location (Mcf/yr)			Average Repair Cost
	Non-Compressor	Reciprocating Compressor	Centrifugal Compressor	
Connections	6.7	-	-	$25
Flanges	88.2	89.7	115	$150
Pressure Relief Valves	3.9	308	-	$150
Other Valves	25	127	63.4	$130
Compressor Seal[a]	-	1,440	485	$2,000
Open-Ended Line (OEL)	43	-	-	$65
Compressor Blowdown OEL	-	1,417	2,887	$5,000

Note: Adapted from exhibit 5 in "Directed Inspection and Maintenance at Gas Processing Plants and Booster Stations" Lessons Learned document. Available online: http://epa.gov/gasstar/documents/ll_dimgasproc.pdf
[a] For the purposes of this paper, compressor seal emissions are not considered leaks.

Table 3-6. Total Average Leak Rate and Repair Costs by Component at Compressor Stations

Component	Average Component Leak Rate by Location (Mcf/yr)		Average Repair Costs	
	On Compressor	Off Compressor	Low	High
Ball/Plug Valves	0.64	5.33	$40	$120
Blowdown Valve	-	207.5	$200	$600
Compressor Valve	4.1	-	$60	$60
Unit Valve	-	3,566	$70	$2,960
Flange	0.81	0.32	$300	$1,250
Open-Ended Line	-	81.8	$45	$45
Pressure Relief Valve	-	57.5	$1,000	$1,000
Connection	0.74	0.6	$10	$30

Note: Adapted from exhibits 4 and 5 in "Directed Inspection and Maintenance at Compressor Stations" Lessons Learned document. Available online: http://epa.gov/gasstar/documents/ll_dimcompstat.pdf.

3.2.5   Update of Fugitive Equipment Leak Emission Factors (CAPP, 2014)

In February of 2014, CAPP issued a report on emission factors for leaks at upstream oil and gas facilities in Alberta and British Columbia. This report served as an update to similar factors that were developed in 2005, prior to the implementation of DI&M BMPs in both these provinces. The report compares the 2005 leak emission factors to the 2014 leak emission factor in order to draw conclusions regarding the effectiveness of the DI&M BMPs in Alberta and British Columbia.

Leak survey results provided by eight industry participants in Alberta and British Columbia were the basis of the emission factors. The results came from 120 facilities and included approximately 276,947 components. All surveys were conducted after 2007. The study authors used this data to develop average emission factors for each type of component and then compared those factors to the factors developed in 2005. Table 3-7 provides a comparison of the emission factors for each type of component from the 2005 study and the 2014 study.

Table 3-7. Comparison of Total Hydrocarbon Leak Emission Factors for Upstream Oil and Gas Facilities that have Implemented DI&M BMPs

Sector	Component	Service[a]	2014 Emission Factor (kg/hour)	2005 Emission Factor (kg/hour)	Ratio of 2014 to 2005 Emission Factors
Gas	Compressor Seal[b]	GV	0.04669	0.71300	0.065
Gas	Connector	GV	0.00082	0.00082	1.000
Gas	Connector	LL	0.00016	0.00055	0.298
Gas	Control Valve	GV	0.03992	0.01620	2.464
Gas	Open-Ended Line	All	0.04663	0.46700	0.100
Gas	Pressure Relief Valve	All	0.00019	0.01700	0.011
Gas	Pump Seal	All	0.00291	0.02320	0.125
Gas	Regulator	All	0.03844	0.00811	4.740
Gas	Valve	GV	0.00057	0.00281	0.205
Gas	Valve	LL	0.00086	0.00352	0.245
Oil	Compressor Seal	GV	0.01474	0.80500	0.018
Oil	Connector	GV	0.00057	0.00246	0.232
Oil	Connector	LL	0.00013	0.00019	0.684
Oil	Control Valve	GV	0.09063	0.01460	6.207
Oil	Open-Ended Line	All	0.15692	0.30800	0.509
Oil	Pressure Relief Valve	All	0.00019	0.01630	0.012
Oil	Pump Seal	All	0.00230	0.02320	0.099
Oil	Regulator	All	0.52829	0.00668	79.085
Oil	Valve	GV	0.00122	0.00151	0.809
Oil	Valve	LL	0.00058	0.00121	0.479

Note: Adapted from Table 10 in "Update of Fugitive Equipment Leak Emission Factors" document (CAPP, 2014). Available online: http://www.capp.ca/getdoc.aspx?DocId=238773&DT=NTV
[a] GV = Gas/Vapor, LL = Light Liquid
[b] For the purposes of this paper, compressor seal emissions are not considered leaks.

The study authors conclude that emissions from leaks have decreased 75% among the survey participants since the implementation of the DI&M programs in Alberta and British Columbia. The leak factors for almost all categories of equipment decreased. The authors did not use this data to develop national or regional estimates of total leak emissions.

4.0 SUMMARY

The EPA has used the information presented in this paper to inform its understanding of leak emissions and potential techniques that can be used to identify and mitigate leaks in the oil and natural gas production, processing, transmission and storage sectors. The following are characteristics the Agency believes are important to understanding this source of VOC and methane emissions:

- The 2014 GHG Inventory estimates there are approximately 332,662 MT of potential methane leak emissions from gas production, 33,681 MT of potential methane leak emissions from gas processing, and 114,348 MT of potential methane leak emissions gas transmission.

- Several studies suggest that the majority of methane and VOC emissions from leaks come from a minority of components (CL, 2013; Clearstone, 2002; and Clearstone, 2006). Furthermore, one study concludes that the majority of methane and VOC emissions from leaks come from a minority of sites (CL, 2013). One study found that the majority of leak emissions from these sites may be attributed to maintenance-related issues such as open thief hatches, failed pressure relief valves, or stuck dump valves (Thoma, 2012).

- The methane and VOC leak emissions from well sites depend on a number of different factors including: the number of wells located at the site, the number of compressors located at the well site and the number and type of processing equipment (separators, heaters, etc.) used at the site.

- Currently, portable analyzers provide an effective approach for both locating and measuring the concentration of leaks from oil and natural gas production sites.

- There are several other technologies being used to detect leaks for the oil and natural gas sectors. These technologies include OGI and ambient/mobile monitoring.

- OGI is being increasingly used to locate leaks in the oil and gas industry. The technology can potentially provide a more time and cost efficient method for locating leaks than traditional technologies, such as portable analyzers. However, there may be limitations to this technology.

    o The technology must be used methodically in order to address certain limitations, such as sensitivities to ambient conditions.

    o OGI technology does not quantify emissions. It may be possible to develop algorithms to quantify emissions with data from OGI, but, to the EPA's knowledge, such algorithms are not currently available.

- Ambient/mobile monitoring and OGI technology might be most effective when used in tandem. In such cases, an OGI inspection could be triggered by the detection of above normal emissions by the ambient/mobile monitoring equipment. This approach potentially could reduce or eliminate OGI inspections at facilities with minimal leak emissions.

- Available information suggests that once a leak is found it is almost always economical to repair the leak. According to the studies reviewed, the cost of detecting the leak is generally far larger than the cost of fixing the leak.

- The CAPP 2014 study and experience through the Natural Gas STAR program suggest DI&M programs can effectively decrease leak emissions.

5.0 CHARGE QUESTIONS FOR REVIEWERS

1. Did this paper appropriately characterize the different studies and data sources that quantify VOC and methane emissions from leaks in the oil and natural gas sector?

2. Please comment on the approaches for quantifying emissions and on the emission factors used in the data sources discussed. Please comment on the national estimates of emissions and emission factors for equipment leaks presented in this paper. Please comment on the activity data used to calculate these emissions, both on the total national and regional equipment counts.

3. Are the emission estimating procedures and leak detection methods presented here equally applicable to both oil and gas production, processing, and transmission and storage sectors?

4. Are there ongoing or planned studies that will substantially improve the current understanding of VOC and methane emissions from leaks and available techniques for detecting those leaks? Please list the additional studies you are aware of.

5. Are there types of wells sites, gathering and boosting stations, processing plants, and transmission and storage stations that are more prone to leaks than others? Some factors that could affect the potential for leaks are the number and types of equipment, the maintenance of that equipment, and the age of the equipment, as well as factors that relate to the local geology. Please discuss these factors and others that you believe to be important.

6. Did this paper capture the full range of technologies available to identify leaks at oil and natural gas facilities?

7. Please comment on the pros and cons of the different leak detection technologies. Please discuss efficacy, cost and feasibility for various applications.

8. Please comment on the prevalence of the use of the different leak detection technologies at oil and gas facilities. Which technologies are the most commonly used? Does the type of facility (e.g., well site versus gathering and boosting station) affect which leak detection technology is used?

9. Please provide information on current frequencies of revisit of existing voluntary leak detection programs in industry and how the costs and emission reductions achieved vary with different frequencies of revisit.

10. Please comment on the potential for using ambient/mobile monitoring technologies in conjunction with OGI technology. This would be a two-phase approach where the ambient/mobile monitoring technology is used to detect the presence of a leak and the OGI technology is used to identify the leaking component. Please discuss efficacy, cost and feasibility.

11. Please comment on the cost of detecting a leak when compared to the cost to repair a leak. Multiple studies described in this paper suggest that detecting leaks is far more costly than repairing leaks and, due to generally low costs of repair and the subsequent product recovery, it is almost always economical to repair leaks once they are found. Please comment on this overall conclusion.

12. If the conclusion is correct that it is almost always economical to repair leaks once they are found, then how important is the quantification of emissions from leaks when implementing a program to detect and repair leaks?

13. Please comment on the state of innovation in leak detection technologies. Are there new technologies under development that are not discussed in this paper? Are there significant advancements being made in the technologies that are not described in this paper?

6.0   REFERENCES

Allen, David, T., et al. 2013. *Measurements of methane emissions at natural gas production sites in the United States.* Proceedings of the National Academy of Sciences (PNAS) 500 Fifth Street, NW NAS 340 Washington, DC 20001 USA. October 29, 2013. 6 pgs. Available at http://www.pnas.org/content/early/2013/09/10/1304880110.full.pdf+html.

Brantley, Halley L., et al. 2014a. *Methane and VOC Emissions from Oil and Gas Production Pads in the DJ Basin.* Forthcoming publication.

Brantley, Halley L., et al. 2014b. *Assessment of Methane Emissions from Oil and Gas Production using Mobile Measurements.* Environmental Science & Technology, Forthcoming publication.

Canadian Association of Petroleum Producers (CAPP). 2007. *Best Management Practices: Management of Fugitive Emissions at Upstream Oil and Gas Facilities.* January 2007. Available at http://www.capp.ca/getdoc.aspx?DocId=116116&DT=PDF.

Canadian Association of Petroleum Producers (CAPP). 2011. *Best Management Practice for Fugitive Emissions Management* (Alberta Energy Regulator, Directive 060: Upstream Petroleum Industry Flaring, Incinerating, and Venting, 8.7. November 3, 2011. Available at http://www.aer.ca/documents/directives/Directive060.pdf.

Canadian Association of Petroleum Producers (CAPP). 2014. *Update of Fugitive Equipment Leak Emission Factors*, February 2014. Available at http://www.capp.ca/getdoc.aspx?DocId=238773&DT=NTV.

Carbon Limits (CL). 2013. *Quantifying cost-effectiveness of systematic Leak Detection and Repair Programs using Infrared cameras.* December 24, 2013. Available at http://www.catf.us/resources/publications/files/CATF-Carbon_Limits_Leaks_Interim_Report.pdf.

Clearstone Engineering Ltd. 2002. *Identification and Evaluation of Opportunities to Reduce Methane Losses at Four Gas Processing Plants.* June 2002.

Clearstone Engineering Ltd. 2006. *Cost-Effective Directed Inspection and Maintenance Control*

Opportunities at Five Gas Processing Plants and Upstream Gathering Compressor Stations and Well Sites. March 2006.

Drilling Information, Inc. (DI). 2011. *DI Desktop.* 2011 Production Information Database.

ERG and Sage Environmental Consulting, LP. 2011. *City of Fort Worth Natural Gas Air Quality Study, Final Report.* Prepared for the City of Fort Worth, Texas. July 13, 2011. Available at http://fortworthtexas.gov/gaswells/default.aspx?id=87074.

EC/R Incorporated. 2011. Memorandum to Bruce Moore, EPA/OAQPS/SPPD from Heather P. Brown, P.E., EC/R Incorporated. *Composition of Natural Gas for Use in the Oil and Natural Gas Sector Rulemaking.* July 28, 2011.

Gas Research Institute (GRI)/U.S. Environmental Protection Agency. 1996. *Research and Development, Methane Emissions from the Natural Gas Industry, Volume 8: Equipment Leaks.* June 1996 (EPA-600/R-96-080h).

ICF Consulting, Inc. 2003. *Identifying Natural Gas Leaks to the Atmosphere with Optical Imaging.* Robinson, D.R. and Luke-Boone, R.E. 2003.

ICF International. *Emissions from Centrifugal Compressors.* 2010.

ICF International. 2014. *Economic Analysis of Methane Emission Reduction Opportunities in the U.S. Onshore Oil and Natural Gas Industries.* ICF International (Prepared for the Environmental Defense Fund). March 2014.

Meister, Mike. 2009. *Smart LDAR – More Cost-Effective?* Environmental Quarterly. Trinity Consultants. July 2009. http://trinityconsultants.com/templates/trinityconsultants/news/article.aspx?id=1293.

Modrak, Mark T., et al. 2012. *Understanding Direct Emissions Measurement Approaches for Upstream Oil and Gas Production Operations.* Air and Waste Management Association 105th Annual Conference and Exhibition, June 19-22, 2012 in San Antonio, Texas.

RTI International. 2011. Memorandum from Cindy Hancy, RTI International, to Jodi Howard, EPA/OAQPS. *Analysis of Emission Reductions Techniques for Equipment Leaks.* December 21, 2011.

RTI International. 2012. Memorandum from Karen Schaffner and Kristin Sroka, RTI International, to Brenda Shine, EPA/OAQPS. *Impacts for Equipment Leaks at Petroleum Refineries.* January 24, 2012.

Thoma, Eben D., et al. 2012. *Assessment of Methane and VOC Emissions from Select Upstream Oil and Gas Production Operations Using Remote Measurements, Interim Report on Recent Studies.* Proceedings of the 105th Annual Conference of the Air and Waste Management Association, June 19-22, 2012 in San Antonio, Texas.

URS Corporation/University of Texas at Austin. 2011. *Natural Gas Industry Methane Emission*

Factor Improvement Study. Final Report. December 2011.
http://www.utexas.edu/research/ceer/GHG/files/FReports/XA_83376101_Final_Report.pdf.

U.S. Energy Information Administration (U.S. EIA). 2012a. Total Energy Annual Energy Review. Table 6.4 Natural Gas Gross Withdrawals and Natural Gas Well Productivity, Selected Years, 1960-2011. (http://www.eia.gov/total energy/data/annual/pdf/sec6_11.pdf).

U.S. Energy Information Administration (U.S. EIA). 2012b. Total Energy Annual Energy Review. Table 5.2 Crude Oil Production and Crude Oil Well Productivity, Selected Years, 1954-2011. (http://www.eia.gov/total energy/data/annual/pdf/sec5_9.pdf).

U.S. Energy Information Administration (U.S. EIA). 2013a. Drilling often Results in both oil and natural gas production. October 2013. Available at http://www.eia.gov/todayinenergy/detail.cfm?id=13571.

U.S. Energy Information Administration (U.S. EIA). 2013b. Annual Energy Outlook 2013. Available at http://www.eia.gov/forecasts/aeo/pdf/0383%282013%29.pdf.

U.S. Environmental Protection Agency (U.S. EPA). 1995. *Protocol for Equipment Leak Emission Estimates.* Office of Air Quality Planning and Standards. Research Triangle Park, NC. November 1995. EPA-453/R-95-017. Available at http://www.epa.gov/ttn/chief/efdocs/equiplks.pdf.

U.S. Environmental Protection Agency Natural Gas STAR Program. 2003a. *Lessons Learned – Directed Inspection and Maintenance at Gate Stations and Surface Facilities.* October 2003. Available at http://www.epa.gov/gasstar/documents/ll_dimgatestat.pdf.

U.S. Environmental Protection Agency Natural Gas STAR Program. 2003b. *Lessons Learned – Directed Inspection and Maintenance at Gas Processing Plants and Booster Stations.* October 2003. Available at http://www.epa.gov/gasstar/documents/ll_dimgasproc.pdf.

U.S. Environmental Protection Agency Natural Gas STAR Program. 2003c. *Lessons Learned – Directed Inspection and Maintenance at Compressor Stations.* October 2003. Available at http://epa.gov/gasstar/documents/ll_dimcompstat.pdf.

U.S. Environmental Protection Agency Natural Gas STAR Program. 2012. *Natural Gas STAR Program Videos.* June 2012. Available at http://www.epa.gov/gasstar/tools/videos.html.

U.S. Environmental Protection Agency (U.S. EPA). 2013. *Petroleum and Natural Gas Systems: 2012 Data Summary. Greenhouse Gas Reporting Program.* October 2013.

U.S. Environmental Protection Agency (U.S. EPA). 2014. *Inventory of Greenhouse Gas Emissions and Sinks: 1990-2012.* Washington, DC. April 2014. Available at http://www.epa.gov/climatechange/Downloads/ghgemissions/US-GHG-Inventory-2014-Chapter-3-Energy.pdf.